World Health Organization. Regional Office for Europe. (2020). *Heated Tobacco Products: A Brief.* World Health Organization. Regional Office for Europe. https://iris.who.int/handle/10665/350470

World Health Organization. Regional Office for Europe. (2020). *Electronic Nicotine and Non-Nicotine Delivery Systems : A Brief.* World Health Organization. Regional Office for Europe. https://iris.who.int/handle/10665/350474

World Health Organization. Regional Office for Europe. (2019). *Country Case Studies On Electronic Nicotine and Non-Nicotine Delivery Systems Regulation: Brazil, Canada, The Republic of Korea and The United Kingdom.* World Health Organization. Regional Office for Europe. https://www.who.int/europe/publications/m/item/country-case-studies-on-electronic-nicotine-and-non-nicotine-delivery-systems-regulation-2019

世界卫生组织（World Health Organization）授权中国科技出版传媒股份有限公司（科学出版社）翻译出版本书中文版。中文版的翻译质量和对原文的忠实性完全由科学出版社负责。当出现中文版与英文版不一致的情况时，应将英文版视作可靠和有约束力的版本。

中文版《新型烟草制品：产品概述及系统监管》
©中国科技出版传媒股份有限公司（科学出版社）2024

世界卫生组织

新型烟草制品

产品概述及系统监管

胡清源　侯宏卫　主译

科学出版社

北　京

内 容 简 介

本书汇编了世界卫生组织有关电子烟产品及不同国家监管案例的3份简报，包括：①加热型烟草制品简报，概述加热型烟草制品类型、释放物成分及对健康的影响；②电子烟碱/非烟碱传输系统简报，对电子烟碱/非烟碱传输系统进行简介，并概述其人群使用情况、成分及健康影响以及对戒烟和起始吸烟所起的作用；③电子烟碱/非烟碱传输系统监管国家案例，以巴西、加拿大、韩国和英国为例，介绍电子烟碱/非烟碱传输系统的监管情况。

本书会引起吸烟与健康、烟草化学和公共卫生学诸多应用领域科学家的兴趣，为客观评价烟草制品的管制和披露措施提供必要的参考。

图书在版编目（CIP）数据

新型烟草制品：产品概述及系统监管 / 世界卫生组织著；胡清源，侯宏卫主译 . -- 北京：科学出版社，2024.6. -- ISBN 978-7-03-078994-5

Ⅰ . TS45

中国国家版本馆 CIP 数据核字第 2024HA4672 号

责任编辑：刘　冉 / 责任校对：杜子昂
责任印制：徐晓晨 / 封面设计：北京图阅盛世

科 学 出 版 社 出版
北京东黄城根北街 16 号
邮政编码：100717
http://www.sciencep.com

北京中石油彩色印刷有限责任公司印刷
科学出版社发行　各地新华书店经销
*

2024 年 6 月第　一　版　　开本：720×1000　1/16
2024 年 6 月第一次印刷　　印张：9
字数：180 000

定价：120.00 元
（如有印装质量问题，我社负责调换）

译 者 名 单

主　译：胡清源　侯宏卫

副主译：陈　欢　李　晓　崔利利　王红娟　田雨闪

译　者：胡清源　侯宏卫　陈　欢　李　晓　崔利利

　　　　　王红娟　田雨闪　韩书磊　付亚宁　刘　彤

目　录

加热型烟草制品简报

摘要 2

致谢 3

引言 4

加热型烟草制品类型 6

加热型烟草制品释放物成分 9

  烟碱传输 9

  主流烟气释放物中的潜在有害物质 9

  侧流烟气和二手烟释放物中的潜在有害物质 10

加热型烟草制品的健康影响 11

  烟碱传输 11

  加热型烟草制品使用者主流烟气暴露的健康风险 11

  加热型烟草制品二手烟暴露的健康风险 11

关键信息 12

结论 13

参考文献 14

电子烟碱/非烟碱传输系统简报

摘要 18

致谢 19

电子烟碱/非烟碱传输系统简介 20

电子烟碱/非烟碱传输系统在人群中的使用 ·················· 21
经常使用电子烟碱/非烟碱传输系统的成年人比例 ·················· 21
经常使用电子烟碱/非烟碱传输系统的青少年比例 ·················· 21
当前使用电子烟碱/非烟碱传输系统的非吸烟青少年比例 ·················· 22

电子烟碱/非烟碱传输系统成分及健康影响 ·················· 23
电子烟碱/非烟碱传输系统气溶胶成分 ·················· 23
电子烟碱/非烟碱传输系统的健康影响 ·················· 24
电子烟碱/非烟碱传输系统气溶胶的二手烟暴露 ·················· 26
呼出气溶胶暴露的健康影响 ·················· 26

电子烟碱/非烟碱传输系统对戒烟和起始吸烟的作用 ·················· 27
电子烟碱/非烟碱传输系统对成年人戒烟的作用 ·················· 27
电子烟碱/非烟碱传输系统对青少年起始吸烟的作用 ·················· 27
调味剂对电子烟碱/非烟碱传输系统使用的作用 ·················· 27

关键信息与结论 ·················· 29
参考文献 ·················· 32

电子烟碱/非烟碱传输系统监管国家案例

摘要 ·················· 38
致谢 ·················· 39
引言 ·················· 40
巴西 ·················· 41
加拿大 ·················· 42
韩国 ·················· 44
英国 ·················· 46
结论 ·················· 50
参考文献 ·················· 50
附录 A 适用于 EN&NNDS 的国家或联邦法规 ·················· 56
附录 B 加拿大各省辖区的其他法规要求 ·················· 67

Contents

Heated Tobacco Products: A Brief

- **Acknowledgements** ·······73
- **Introduction** ·······74
- **Types of HTPs** ·······75
- **HTP emission content** ·······78
  - Nicotine delivery ·······78
  - Potentially toxic substances in mainstream emission ·······78
  - Potentially toxic substances in side-stream and second-hand emission ·······79
- **Effects of HTP use on health** ·······80
  - Nicotine delivery ·······80
  - Health risks to HTP users from exposure to mainstream emission ·······80
  - Health risks from exposure to HTP second-hand emission ·······80
- **Key messages** ·······81
- **Conclusions** ·······82
- **References** ·······83

Electronic Nicotine and Non-Nicotine Delivery Systems: A Brief

- **Acknowledgements** ·······87
- **Electronic nicotine and non-nicotine delivery systems** ·······88
- **EN&NNDS use among the population** ·······89
  - Proportion of the adult population using EN&NNDS regularly ·······89
  - Proportion of young people using EN&NNDS regularly ·······89

  EN&NNDS current use among non-smoking young people ·····90
EN&NNDS contents and health effects ·····91
  EN&NNDS aerosol contents ·····91
  Health effects of using EN&NNDS ·····92
  Second-hand exposure to EN&NNDS aerosol ·····94
  Health effects of exposure to exhaled aerosol ·····94
EN&NNDS' role in smoking cessation and initiation ·····95
  EN&NNDS' role in smoking cessation among adults ·····95
  EN&NNDS' role in smoking initiation among young people ·····95
  The role of flavours in EN&NNDS initiation and use ·····95
Key messages and conclusions ·····97
References ·····99

Country Case Studies On Electronic Nicotine and Non-Nicotine Delivery Systems Regulation, 2019: Brazil, Canada, The Republic of Korea and The United Kingdom

Acknowledgements ·····105
Introduction ·····106
Brazil ·····107
Canada ·····108
The Republic of Korea ·····110
The United Kingdom ·····112
Conclusions ·····115
References ·····116
Annex 1 National or federal regulation that applies to EN&NNDS ·····120
Annex 2 Additional regulatory requirements by the provincial jurisdictions of
     Canada ·····132

加热型烟草制品简报

摘　　要

加热型烟草制品（HTP）是能产生含有烟碱和其他化学物质的释放物并由使用者吸入体内的烟草制品。HTP是一类新出现的烟草制品，以所谓的潜在暴露降低甚至是风险改良的烟草制品的形式销售。目前尚无足够的证据可以得出HTP的危害性低于传统卷烟的结论。实际上，令人担忧的是，尽管与传统卷烟相比，HTP使用者某些有害物质的暴露水平更低，但其他有害物质的暴露水平更高。目前尚不清楚这种毒理学特征如何转化为短期和长期的健康影响。世界卫生组织《烟草控制框架公约》（WHO FCTC）缔约方大会认定HTP为烟草制品，因此认为其应遵守WHO FCTC的规定。

关键词：加热型烟草制品；释放物；健康影响；烟草；世界卫生组织《烟草控制框架公约》；监管

致　　谢

这份简报由世界卫生组织欧洲地区办事处顾问Armando Peruga主持编写，世界卫生组织总部科学家Ranti Fayokun、项目经理Kristina Mauer-Stender、技术官员Angela Ciobanu及世界卫生组织欧洲地区办事处非传染性疾病和终生健康促进司烟草控制计划顾问Elizaveta Lebedeva参与编写。

作者还要感谢世界卫生组织欧洲地区办事处非传染性疾病和终生历程健康促进司司长Bente Mikkelsen对本简报的编写工作给予总体指导和支持。

本项目由德国政府提供资金。

引　言

加热型烟草制品（HTP）[1]中加工过的烟草在未被点燃的情况下被加热，产生了含有烟碱和其他化学物质的释放物，供使用者吸入。HTP是一类新出现的烟草制品，以所谓的潜在暴露降低甚至是风险改良的烟草制品的形式销售。此类产品被定义为"新型产品"，因为在作此简报之时，HTP从概念上和技术上都是从烟草公司20世纪80~90年代销售的类似产品发展而来的。当时，这些产品的前体产品并不成功，因此停售。但是，新兴的HTP有望占据重要的市场份额。

2016年HTP的总销售额为21亿美元，预计到2021年将达到179亿美元[1]。HTP现如今有更好的获利营销机会，因为烟草行业在一定程度上依赖电子烟碱/非烟碱传输系统（EN&NNDS）在一些国家的普及。尽管是完全不同的产品类别，但在许多国家，EN&NNDS改变了人们对抽吸传统卷烟和使用烟碱传输装置的社会规范和观念。

到目前为止，仅有几种HTP在市场上销售[1]。日本烟草国际公司（JTI）于2013年与一家名为Pax Labs的公司联合推出了Ploom，该公司继续独立销售PAX品牌。JTI于2016年独立重新推出了Ploom。菲利普·莫里斯国际公司（PMI）于2014年推出了IQOS（"我戒掉了普通的烟"）。英美烟草公司（BAT）于2015年首先在罗马尼亚上市了iFuse。后来，BAT在亚洲上市了Glo。韩国烟草人参公司（KT&G）于2017年推出了lil，成为最新进入HTP市场的企业[2]。当前，HTP在大约40个国家/地区销售，并且IQOS出现在其中大多数国家/地区。

关于HTP使用率的信息不多，有关其趋势的信息更少。在日本，2015年15~69岁年龄段人群中有0.3%报告在过去30天内使用过IQOS（当前使用）[3]。两年后，这一数字为3.6%。2017年，有1.2%的人正在使用Ploom，0.8%的人正在使用Glo[4]。这些数字不是互斥的。在意大利，2017年15岁以上人群中有1.4%尝试了IQOS。总体上，1.0%的从未吸烟者、0.8%的既往吸烟者和

1　烟草行业也将HTP称为"加热不燃烧型产品"。

3.1%的当前吸烟者尝试了IQOS[5]。在德国，2017年有0.3%年龄在14岁以上的当前吸烟者和最近戒烟者正在使用HTP[6]。在英国，2017年有1.7%的成年人尝试过或正在使用HTP，但其中只有13%每天使用[7]。2017年在韩国引入IQOS三个月后，19~24岁的年轻人中有3.5%是当前使用者，尽管他们都还使用了传统卷烟和EN＆NNDS[8]。

吸烟是通过燃烧烟草来摄取烟碱的传统方法，烟气中含有成千上万种化合物，其中许多对健康有害。HTP技术基于这样的原理，即燃烧烟草对于释放烟碱不是必要的。在吸烟中，烟碱的雾化是通过点燃烟草实现的，燃烧锥的温度最高达900℃，但是HTP在约350℃的温度下通过烟草的挥发甚至热解，也可以实现类似的释放[9]，尽管在某些产品中温度可能会达到550℃[10]。烟碱在较低温度下挥发，与传统卷烟相比，使用者接触到的释放物中有害物质的种类更少，含量也更低。HTP和EN＆NNDS之间的本质区别在于，前者使用烟叶，而后者不使用。

本部分概述了有关HTP的成分、释放物和健康影响的现有证据，并对监管政策的选择进行了审查。

加热型烟草制品类型

HTP按照如何加热烟草将烟碱输送到使用者的肺部分为四种类型[11]。第一种类型是一种类似卷烟的装置，内嵌热源，可用于雾化烟碱。热量由位于产品末端的炭头提供，必须像传统卷烟一样用标准火柴或打火机点燃（图1-1）。点燃后，热量从炭头端传递到烟草，两者是不接触的。由此产生约350℃的温度，含烟碱的释放物通过烟嘴吸入。没有使用电气系统。使用后，需要将产品熄灭并丢弃[12]。

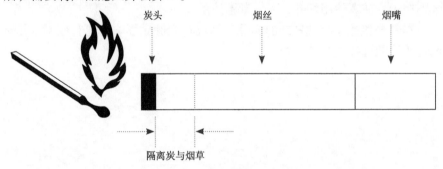

图1-1　HTP第一种类型

第二种类型使用外部热源将专门设计的卷烟中的烟碱雾化。这是IQOS[13]（图1-2）和Glo[14]的基本设计。PMI的HTP中使用的烟草显然不是典型的烟丝，而是一种强化的重构的网状烟叶（一种再造烟草），其中5%~30%（质量分数）的化合物会形成释放物，例如多元醇、乙二醇酯和脂肪酸。在IQOS中，烟草通过插入加热棒（或包含烟草的元件）末端的加热装置中的页片加热，在抽吸时热量通过烟嘴上的烟草塞散发。然后，释放物通过中空的醋酸纤维素管和聚合物薄膜过滤器到达口腔。BAT将其Glo产品描述为一个加热管，由两个单独控制的腔室组成，这些腔室由设备上的一个按钮激活，在30~40 s内达到工作温度（240℃）。

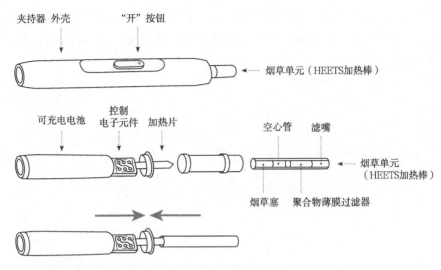

图1-2　HTP第二种类型

第三种使用类似微型烤箱的加热密封腔（图1-3）。电池为加热室提供能量，加热室通过物理接触将热量传递给使用者置于其中的任何材料。使用者必须用磨碎的烟叶填充微型加热室以雾化烟碱，然后通过烟嘴吸入释放物。这就是干药草或松散烟叶的雾化器（例如Pax）的工作方式[15]。与其他HTP不同的是，制造商不提供也不建议任何材料填充液体插件腔室。

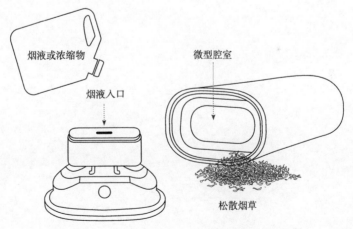

图1-3　HTP第三种类型

第四种使用类似于EN&NNDS的技术，从少量烟草中提取风味成分。

英美烟草公司的iFuse产品[16]似乎是一种混合的ENDS烟草产品,其释放物通过并加热烟草以吸收风味物质,然后被使用者吸入。JTI Ploom TECH以类似的方式运行[17]。

加热型烟草制品释放物成分

烟碱传输

IQOS的主流释放物似乎比每支传统卷烟传输的烟碱更少。在研究中，主流释放物的烟碱含量为参比卷烟的57%~83%。Glo和iFuse的烟碱含量低于IQOS（为传统卷烟的19%~23%）。与早期的ENDS相比，HTP释放出更多的烟碱[18]，但无法获得与第三代ENDS的比较。烟草行业和独立资助的研究中烟碱释放量的测量结果相似[18]。

对人类使用HTP后血浆烟碱水平进行测量的研究表明，HTP的烟碱递送量因品牌而异，但除IQOS以外，都低于传统卷烟。HTP传递的烟碱在血浆中的浓度达到峰值的速度与传统卷烟一样快[18]。

主流烟气释放物中的潜在有害物质

HTP释放物所含有害和潜在有害化合物（HPHC）的种类与传统卷烟烟气几乎相同，尽管在某些情况下含量较低。对已发表的同行评审文章的系统审查表明，所分析的有害物质水平比卷烟烟气至少低62%，颗粒物（PM）比卷烟烟气低75%[18]。烟草行业和独立资助的研究，包括德国[19]、荷兰[20]和英国[21]的一些政府机构，发现HTP释放物中的有害物质含量低于卷烟烟气。不过，与烟草行业资助的研究相比，独立研究报告的焦油量更低，但烟草特有亚硝胺以及乙醛、丙烯醛和甲醛的含量更高[18]。

关于HTP释放物中有害物质含量更低的结论，必须注意以下几点：

迄今为止，同行评议文章中测量的有害物质种类并未涵盖所有相关的HPHC。例如，PMI在其向美国食品药品管理局（FDA）提交的材料中报告了IQOS主流释放物中FDA建议的93种HPHC中的40种的含量。未报告的53种HPHC的水平未知，其中50种是致癌的[20]。

PMI向FDA提交的报告包括57种未列入FDA HPHC清单的其他成分含

量。IQOS释放物中有56种的水平高于传统卷烟。其中22种化合物的含量是参比卷烟中的2倍，7种化合物的含量比参比卷烟中高10倍以上。IQOS似乎降低了某些有害物质的暴露，但增加了其他物质的暴露。其中一些物质属于已知具有重大毒性的化学类别，但总的来说，其中许多物质的毒性信息有限[22]。

侧流烟气和二手烟释放物中的潜在有害物质

与传统卷烟一样，但与EN＆NNDS不同的是，HTP会产生侧流释放物。三项研究（一项独立资助，两项与烟草行业有关联）报告了IQOS和Glo中某些HPHC的含量。这些研究都发现，二手烟中存在甲醛和乙醛，尽管其含量分别比卷烟烟气中低约10~20倍。只有独立研究发现二手烟中有颗粒物（PM）和丙烯醛。在这项研究中，PM大约比卷烟烟气中低4倍，而丙烯醛则低约50倍[18]。因此，有证据表明，HTP的二手烟释放物使非吸烟者暴露于可量化水平的PM和主要有害物质，但其水平低于燃烧型烟草制品。

加热型烟草制品的健康影响

烟碱传输

一些HTP（但不是全部），尤其是IQOS的烟碱传递特性与传统卷烟相似。因此，尽管据报道使用者满意度低于传统烟草制品，但某些HTP可能是卷烟传输烟碱的适当替代品。

加热型烟草制品使用者主流烟气暴露的健康风险

目前尚无证据可以断定使用HTP是否与暴露于主流释放物的任何长期临床结果（积极或消极）有关。PMI一项研究声称，与抽吸传统卷烟相比，IQOS减少了与内皮功能障碍、氧化应激、炎症以及高密度脂蛋白和胆固醇计数相关的生物标志物[23]。PMI还在提交给美国FDA的文件中声称："人类临床研究已证实……炎症的临床标志物显示出积极的变化，类似于戒烟后出现的变化。"然而，对PMI数据进行严格审查得出的结论是，PMI没有提供直接的人类肺部临床数据。审查还得出结论，在使用者中，没有证据表明改用IQOS的吸烟者的肺部炎症或肺功能得到了改善。因此，PMI声称改用IQOS的吸烟者可以减少肺部炎症和慢性阻塞性肺疾病的风险，甚至连他们自己的数据也无法证明。只有极少数研究报告了使用HTP的短期影响。这些研究表明存在某些短期的生理病理效应[24-26]。

加热型烟草制品二手烟暴露的健康风险

目前尚无证据表明使用HTP是否与暴露于二手烟的任何长期临床结果相关。然而，HTP仍会产生带有超微颗粒和一些有害物质的侧流释放物，尽管其含量低于传统卷烟。最近的一项研究发现，一部分暴露于IQOS二手释放物的人出现了短期症状，如喉咙痛、眼痛和身体不适[4]。

鉴于包括世界卫生组织[27,28]的一些公共卫生组织认为任何水平的侧流暴露都不是安全的或可以接受的，这些发现显然令人担忧，值得进一步研究。

关 键 信 息

HTP含有烟草并释放烟碱和其他有害物质。

HTP产生主流释放物和侧流释放物。吸入主流释放物会使HTP使用者暴露于其中所含的有害物质。非吸烟者可能会吸入侧流释放物或二手烟。

目前没有足够的证据得出HTP的危害性低于传统卷烟的结论。实际上，令人担忧的是，尽管与传统卷烟相比，HTP使用者某些有害物质的暴露水平更低，但其他有害物质的暴露水平更高。目前尚不清楚这种毒理学特征如何转化为短期和长期的健康影响。

结　　论

各国政府应引入新型烟草制品（包括HTP）上市前评估制度。除非有确凿的证据表明与传统卷烟相比，HTP可减少有害和潜在有害成分的暴露并降低健康风险，否则不应允许销售该产品。

如果政府无法阻止HTP进入市场，或在没有此类证据的情况下决定允许HTP销售，则应确保烟草行业不能将政府对该产品的授权等同于其认可。

此外，依据世界卫生组织《烟草控制框架公约》（WHO FCTC）第八次缔约方大会FCTC/COP8（22）号决议中的建议[29]，HTP应与其他烟草制品一样征税。该决议将HTP视为烟草制品，因此认为其应遵守WHO FCTC的规定。该决议还提醒各缔约方根据WHO FCTC和国家法律优先采取以下措施：

- 防止非吸烟者开始使用HTP；
- 根据WHO FCTC第8条，保护人们免受HTP释放物的危害，并明确将无烟立法的范围扩大到这些产品；
- 防止对HTP做出健康声明；
- 根据WHO FCTC第13条，采取有关HTP的广告、促销和赞助的措施；
- 根据WHO FCTC第9条和第10条，对HTP进行成分管制和披露；
- 根据WHO FCTC第5.3条，保护烟草控制政策和活动不受与HTP相关的所有商业利益和其他既得利益的影响，包括烟草行业的利益；
- 根据国家法律对HTP的生产、进口、分销、展示、销售和使用进行监管，包括酌情限制或禁止，同时考虑对人体健康的高度保护。

最后，重要的是，不仅要全面监测市场发展情况，而且要通过在所有适当的调查中纳入相关问题，全面监测HTP的使用情况。

参考文献[1]

[1] Heated tobacco products (HTPs) market monitoring information sheet. In: World Health Organization [website]. Geneva: World Health Organization; 2019 (https://www.who.int/tobacco/publications/prod_regulation/htps-marketing-monitoring/en/).

[2] Lee J, Lee S. Korean-made heated tobacco product, ilil"ilTob Control 2018:tobaccocontrol-2018-054430. doi:10.1136/tobaccocontrol-2018-054430.

[3] Tabuchi T, Kiyohara K, Hoshino T, Bekki K, Inaba Y, Kunugita N. Awareness and use of electronic cigarettes and heat-not-burn tobacco products in Japan. Addiction 2016;111(4):706–13. doi:10.1111/add.13231.

[4] Tabuchi T, Gallus S, Shinozaki T, Nakaya T, Kunugita N, Colwell B. Heat-not-burn tobacco product use in Japan: its prevalence, predictors and perceived symptoms from exposure to secondhand heat-not-burn tobacco aerosol. Tob Control 2017;27(e1):e25–33. doi:10.1136/tobaccocontrol-2017-053947.

[5] Liu X, Lugo A, Spizzichino L, Tabuchi T, Pacifici R, Gallus S. Heat-not-burn tobacco products: concerns from the Italian experience. Tob Control 2018;tobaccocontrol-2017-054054. doi:10.1136/tobaccocontrol-2017-054054.

[6] Kotz D, Kastaun S. E-Zigaretten und Tabakerhitzer: reprprrhitzer Daten zu Konsumverhalten und assoziierten Faktoren in der deutschen Bevvschen (die DEBRA-Studie) [E-cigarettes and tobacco heaters: representative data on consumer behaviour and associated factors in the German population (the DEBRA study)]. Bundesgesundheitsblatt Gesundheitsforschung Gesundheitsschutz 2018;61(11):1407–14. doi:10.1007/s00103-018-2827-7 [in German].

[7] Brose L, Simonavicius E, Cheeseman H. Awareness and use of "heat-not-burn" tobacco products in Great Britain. Tob Regul Sci. 2018;4(2):44–50. doi:10.18001/trs.4.2.4.

[8] Kim J, Yu H, Lee S, Paek Y. Awareness, experience and prevalence of heated tobacco product, IQOS, among young Korean adults. Tob Control 2018;27(Suppl. 1):s74–7. doi:10.1136/tobaccocontrol-2018-054390.

[9] Davis B, Williams M, Talbot P. iQOS: evidence of pyrolysis and release of a toxicant from plastic. Tob Control 2019;28:34–41. doi:10.1136/tobaccocontrol-2017-054104.

[10] Jiang Z, Ding X, Fang T, Huang H, Zhou W, Sun Q. Study on heat transfer process of a heat not burn tobacco product flow field. J Phys Conf Ser. 2018;1064:012011. doi:10.1088/1742-6596/1064/1/012011.

[11] O'Connor R. Heated tobacco products. In: WHO study group on tobacco product regulation. Report on the scientific basis of tobacco product regulation: seventh report of a WHO study group. Geneva: World Health Organization; 2019:3–29 (https://apps.who.

---

1 所有网络链接均于 2019 年 12 月 1 日访问。

[12] Platform 2. In: PMI Science [website]. Neuchatel: PMI Science; undated (https://www.pmiscience.com/our-products/platform2).

[13] Our tobacco heating system: IQOS. In: Philip Morris International [website]. Neuchatel: Philip Morris International; undated (https://www.pmi.com/smoke-free-products/iqos-our-tobaccoheating-system).

[14] Tobacco heating products. In: British American Tobacco. London: British American Tobacco; 2019 (https://www.bat.com/group/sites/UK__9D9KCY.nsf/vwPagesWebLive/DOAWUGNJ#).

[15] PAX 3 FAQ. In: Pax Labs [website]. San Francisco (CA): Pax Labs; undated (https://www.pax.com/pages/pax-3-faq).

[16] Spencer B. The iFuse Fuse encer B. The iFuse bs [website]. San Francisco (CA): Pax Labs; undatflavour of the vapour. Daily Mail online. 23 November 2013 (https://www.dailymail.co.uk/sciencetech/article-3330238/The-iFuse-hybrid-cigarette-combines-e-cig-technology-tobaccoimprove-flavour-vapour.html).

[17] Reduced-risk products (RRP). In: Japan Tobacco International [website]. Geneva: Japan Tobacco International; undated (https://www.jt.com/sustainability/our_business/tobacco/rrp/index.html).

[18] Simonavicius E, McNeill A, Shahab L, Brose L. Heat-not-burn tobacco products: a systematic literature review. Tob Control 2018; tobaccocontrol-2018-054419. doi:10.1136/tobaccocontrol-2018-054419.

[19] Mallock N, B E, McNeill R, Danziger M, Welsch T, Hahn H et al. Levels of selected analytes in the emissions of ons of of tobaccocontrol-2018-054419. doi:10.1136/tobaccocontrol-2ccocontrol-2ol-2tmToxicol. 2018;92(6):2145–9. doi:10.1007/s00204-018-2215-y.

[20] Addictive nicotine and harmful substances also present in heated tobacco. In: National Institute for Public Health and the Environment (RIVM) [website]. Bilthoven: National Institute for Public Health and the Environment (RIVM); 2018 (https://www.rivm.nl/en/news/addictivenicotine-and-harmful-substances-also-present-in-heated-tobacco).

[21] Statement on heat not burn tobacco products. London: Food Standards Agency; 2017 (https://cot.food.gov.uk/committee/committee-on-toxicity/cotstatements/cotstatement-syrs/cotstatements-2017/statement-on-heat-not-burn-tobacco-products).

[22] St Helen G, Jacob Iii P, Nardone N, Benowitz N. IQOS: examination of Philip Morris Internationalnationalonalationalf Philip MoTob Control 2018;27(Suppl. 1):s30–6. doi:10.1136/tobaccocontrol-2018-054321.

[23] L3. Li: F, Picavet P, Baker G, Haziza C, Poux V, Lama N et al. Effects of switching to thementhol tobacco heating system 2.2, smoking abstinence, or continued cigarette smoking on clinically relevant risk markers: a randomized, controlled, open-label, multicenter study in sequential confinement and ambulatory settings (part 2). Nicotine Tob Res. 2017;20(2):173173 2.2, smoking ntr/ntx028.

[24] Leigh N, Tran P, O'Connor R, Goniewicz M. Cytotoxic effects of heated tobacco products (HTP) on human bronchial epithelial cells. Tob Control 2018;27(Suppl. 1):s26–9. doi:10.1136/tobaccocontrol-2018-054317.

[25] Biondi-Zoccai G, Sciarretta S, Bullen C, Nocella C, Violi F, Loffredo L et al. Acute effects of heat-not-burn, electronic vaping, and traditional tobacco combustion cigarettes: the Sapienza University of Rome vascular assessment of proatherosclerotic effects of smoking (SUR-VAPES 2 randomized trial. J Am Heart Assoc. 2019;8(6): e010455. doi:10.1161/jaha.118.010455.

[26] Sohal S, Eapen M, Naidu V, Sharma P. IQOS exposure impairs human airway cell homeostasis: direct comparison with traditional cigarette and e-cigarette. ERJ Open Res. 2019;5(1):00159-2018. doi:10.1183/23120541.00159-2018.

[27] Policy recommendations on protection from exposure to second-hand tobacco smoke. Geneva: World Health Organization; 2007 (https://www.who.int/tobacco/publications/second_hand/protection_second_hand_smoke/en/).

[28] Guidelines for implementation of Article 8. Guidelines on the protection from exposure to tobacco smoke. Geneva: World Health Organization; 2007 (https://www.who.int/fctc/treaty_instruments/adopted/article_8/en/).

[29] Conference of the Parties to the WHO Framework Convention on Tobacco Control. Eighth session. Geneva, Switzerland, 1–6 October 2018. Decision. FCTC/COP8(22). Novel and emerging tobacco products. Geneva: World Health Organization; 2018 (https://www.who.int/fctc/cop/sessions/cop8/FCTC__COP8(22).pdf).

电子烟碱/非烟碱传输系统简报

摘　　要

电子烟碱/非烟碱传输系统（EN&NNDS）是一类使用电加热线圈将电子烟烟液转化为气溶胶供使用者吸入的异质性产品。EN&NNDS并非无害。尽管尚未充分研究其对发病率和死亡率的长期影响，但EN&NNDS对于青少年、孕妇和从未吸烟的成年人并不安全。虽然预计这些人群使用EN&NNDS可能会增加其健康风险，但成年吸烟者（不包括孕妇）完全并迅速地从燃烧型卷烟转换为仅使用无掺杂的且经过适当监管的EN&NNDS，则可能会降低健康风险。决定监管EN&NNDS的成员国可以着重考虑以下几方面：对以医药产品或治疗设备做出健康声明的EN&NNDS进行监管；禁止或限制EN&NNDS的广告、促销和赞助；通过禁止在所有室内场所或禁止吸烟的场所使用EN&NNDS来最大限度地降低非使用者的健康风险；限制在EN&NNDS中允许使用的特定调味剂的含量和种类，以减少其吸引年轻人开始使用EN&NNDS的情况。

关键词：电子烟碱传输系统（ENDS）；电子非烟碱传输系统（ENNDS）；EN&NNDS成分；健康影响；对戒烟的作用；世界卫生组织《烟草控制框架公约》；监管

致　　谢

　　这份简报由世界卫生组织欧洲地区办事处顾问Armando Peruga主持编写，世界卫生组织总部科学家Ranti Fayokun，项目经理Kristina Mauer-Stender，技术官员Angela Ciobanu及世界卫生组织欧洲地区办事处非传染性疾病司和终生健康促进司烟草控制计划顾问Elizaveta Lebedeva参与编写。

　　作者还要感谢世界卫生组织欧洲地区办事处非传染性疾病和终生健康促进司司长Bente Mikkelsen对本简报的编写工作给予总体指导和支持。

　　本项目由德国政府提供资金。

电子烟碱／非烟碱传输系统简介

电子烟碱/非烟碱传输系统（EN&NNDS）[1]是一类使用电加热并将液体转变成气溶胶供使用者吸入的异质性产品。

EN&NNDS使用或雾化过程气溶胶的产生及其组成以及随后的气溶胶物质暴露取决于以下四个因素：

（1）电子烟烟液组成；

（2）用于制造装置的材料；

（3）用于操作加热电子烟烟液的电力或功率；

（4）使用EN&NNDS时的抽吸曲线或吸入特性。

电子烟烟液都包含载液（保润剂），占烟液体积的80%~90%，还有一些水（占烟液体积的10%~20%），通常还有烟碱和调味剂。电子烟烟液中使用的主要基质丙二醇和甘油在与雾化器加热线圈接触时部分分解形成多种有害物质，包括羰基化合物。电子烟烟液也可能含有烟碱（一种高度成瘾的物质），可能对胎儿和青少年的大脑发育产生不良影响。

EN&NNDS中的加热元件或线圈通常由各种金属（如镍）或金属合金（如镍铬合金）的电阻丝制成。设备的金属部件有时会用铅焊接。

为了加热和雾化电子烟烟液，当EN&NNDS激活时，来自电池的电流流经线圈。达到的温度取决于产生的电能，而电能又取决于电池提供的能量和线圈的电阻。电阻越低，通过的电流就越大，线圈产生的热量就越多。在正常操作条件下，电子烟烟液的温度在100~350℃。

使用者的吸入行为或抽吸曲线具有以下变量：抽吸容量、吸入深度、抽吸速率和抽吸强度。这些变量决定了吸入气溶胶的量以及气溶胶进入呼吸系统的深度。

1　本简报遵循世界卫生组织《烟草控制框架公约》缔约方大会的术语,电子烟碱传输系统（ENDS）和电子非烟碱传输系统（ENNDS）分别指烟液中含和不含烟碱的电子烟产品。这些系统统称为电子烟碱／非烟碱传输系统（EN&NNDS），被普遍称为电子烟。其他来源将 ENDS 称为替代型烟碱传输系统（ANDS）。

电子烟碱/非烟碱传输系统在人群中的使用

经常使用电子烟碱/非烟碱传输系统的成年人比例

在美国和欧盟这两个世界上主要的电子烟市场，当前使用EN&NNDS（定义为在过去30天至少使用一次）的成年人比例在2018年和2017年分别为3.2%[1]和2%[2]。2018年，欧盟国家中使用率最高的是英国（英格兰），为6.2%[3]。在新西兰，2017~2018年间，有3.8%的成年人当前使用EN&NNDS[4]。来自9个国家/地区的其他数据表明，2017~2018年间，大多数国家/地区有不超过4%的成年人经常使用EN&NNDS[5]。

很少有国家提供发展趋势数据。自2014年以来，美国当前使用EN&NNDS的成年人比例一直稳定保持在3.7%[1]，而自2015年以来欧盟保持在2%[6]。在加拿大，2013~2017年间，过去30天使用和每日使用EN&NNDS的成年人比例保持稳定[7]。仅新西兰显示成年人中EN&NNDS的使用率明显提升，从2015~2016年的1.4%上升至2017~2018年的3.8%和2018~2019年的4.7%[4]。大多数EN&NNDS使用者是当前或既往吸烟者。

经常使用电子烟碱/非烟碱传输系统的青少年比例

来自22个国家的13~15岁青少年当前使用EN&NNDS的数据表明，经常使用EN&NNDS的青少年比例高于成年人。2017~2019年间，青少年的使用率从日本的0.7%到乌克兰的18.4%不等，各国中位数为8.1%[8]。

2008~2015年间，波兰、新西兰、韩国和美国的青少年既往使用EN&NNDS的比例有所增加，加拿大和意大利有所下降，而英国则保持稳定[9]。2017~2018年间，美国11~18岁青少年当前使用EN&NNDS的人数有所增加[10]，而英国则保持稳定。2019年，英国有1.6%的11~18岁青少年每周使用EN&NNDS超过一次，而2018年这一比例为1.7%[11]。在加拿大，

2016~2017年间，7~9年级的青少年过去30天使用EN&NNDS的比例为5.4%，与2014~2015年间的使用率相比没有显著差异[12]。最近的一项研究比较了2017~2018年间加拿大、英国（英格兰）和美国16~19岁青少年使用EN&NNDS的变化情况，证实了加拿大和美国过去30天和过去1周内使用EN&NNDS的情况有所增加，而英国（英格兰）则保持稳定（表2-1）[13]。

表2-1　2017~2018年间三个国家16~19岁青少年当前使用EN&NNDS的流行率变化

使用EN&NNDS	加拿大		美国		英国（英格兰）	
	2017（%）	2018（%）	2017（%）	2018（%）	2017（%）	2018（%）
过去30天	8.4	14.6	11.1	16.2	8.7	8.9
过去1周	5.2	9.3	6.4	10.6	4.6	4.6

资料来源：Hammond等[13]

当前使用电子烟碱/非烟碱传输系统的非吸烟青少年比例

美国的数据显示，2017年，从未吸烟的11~18岁青少年中有0.8%经常使用EN&NNDS（过去10天内至少使用一次）。在2018年，这一比例上升至2.4%[14]。而在英国（英格兰），从未吸烟的11~18岁青少年中有0.8%当前使用EN&NNDS[3]。17~18岁从未吸烟者每周使用EN&NNDS的比例在2016年和2017年为0%[15]，2018年为0.2%[3]。最近的一项研究比较了2017~2018年间加拿大、英国（英格兰）和美国的16~19岁青少年使用EN&NNDS的变化情况，证实了加拿大和美国过去30天和过去1周内从未吸烟者使用EN&NNDS的人数不断增加，而在英国（英格兰）这一数字没有变化（表2-2）[13]。

表2-2　2017~2018年间三个国家16~19岁从不吸烟者当前使用EN&NNDS的流行率变化

使用EN&NNDS	加拿大		美国		英国（英格兰）	
	2017（%）	2018（%）	2017（%）	2018（%）	2017（%）	2018（%）
过去30天	2.3	5.0	2.4	5.9	1.6	1.9
过去1周	0.8	2.7	1.1	3.0	0.5	0.4

资料来源：Hammond等[13]

电子烟碱/非烟碱传输系统成分及健康影响

电子烟碱/非烟碱传输系统气溶胶成分

使用者从EN&NNDS吸入的气溶胶，除包含电子烟烟液中的烟碱外，还含有许多潜在有害物质。EN&NNDS释放的气溶胶中潜在有害物质的种类、含量和特性变化很大，并且取决于产品特性（包括设备和电子烟烟液特性）以及使用者如何操作设备。但是，在典型的使用条件下，除某些金属外，不含杂质的EN&NNDS所释放的潜在有害物质的种类和浓度要低于烟草烟气。

气溶胶中引起健康问题的主要物质是金属，例如铬、镍和铅，以及羰基化合物，例如甲醛、乙醛、丙烯醛和乙二醛。

金属的种类和浓度取决于产品特征和使用时的吸入模式。某些水平的金属暴露可能会对健康造成严重影响，例如神经、心血管和呼吸系统疾病。气溶胶中的金属种类可能比燃烧型卷烟中的多，并且在某些情况下，其浓度比卷烟烟气中的高。金属有可能来自用于加热电子烟烟液的金属线圈和设备的焊接接头。通过适当的装置设计，可以在很大程度上防止金属释放。

羰基化合物对使用者有潜在的危害。甲醛是人体致癌物，乙醛可能对人体致癌，丙烯醛是呼吸系统的强烈刺激物，乙二醛显示出致突变性。大多数羰基化合物来自保润剂丙二醇和甘油的热分解。气溶胶中检测到的羰基种类和含量低于燃烧型烟草烟气，但即使是这些含量水平也会引起健康问题。

可能对健康造成影响的气溶胶中的其他物质是颗粒物和某些调味剂。EN&NNDS气溶胶中的颗粒物数量和大小与燃烧型烟草烟气中的颗粒物数量和大小没有太大差异。然而，颗粒物的组成并不相同，并且可能对健康产生不同的影响。来自EN&NNDS的气溶胶颗粒物主要由水滴和保润剂液滴组成，而燃烧型烟草烟气中的颗粒物主要是复杂的有机成分，其中含有已知或可疑的致癌物。

因此，尽管EN&NNDS产生的颗粒物会对健康造成影响，但与烟草烟气中的颗粒物相比，预计其对健康的危害较小。

在加热和吸入时，某些调味剂（例如二乙酰、肉桂醛和苯甲醛）被认为是引起健康问题的原因。当电子烟烟液中含有烟碱时，气溶胶中也含有烟碱。ENDS使用者吸入的烟碱的量变化很大，并且取决于产品特性（包括设备和电子烟烟液特性）以及设备的操作方式。有大量证据表明，有经验的成年ENDS使用者从ENDS摄入烟碱的量与从燃烧型卷烟吸入的烟碱量相当。

电子烟碱 / 非烟碱传输系统的健康影响

科学家仍在了解EN&NNDS对健康的长期影响。当前，尚无足够的研究来确定使用无掺杂的且经过适当监管的EN&NNDS是否与心血管、肺部或癌症有关。

以下为美国国家科学院、工程院和医学院（NASEM）在2018年按证据强度评估的目前已知情况[16][1]。

有确凿证据表明：

- 使用EN&NNDS完全替代燃烧型卷烟可以减少使用者暴露于燃烧型卷烟中的多种有害物质和致癌物质；
- 如果电池质量不佳、储存不当或被使用者改装，则EN&NNDS装置可能爆炸，造成灼伤和弹射伤害；
- 有意或无意接触电子烟烟液（饮用、眼睛接触或皮肤接触）可导致不良健康影响，甚至致命。

有充分证据表明：

- ENDS的使用会导致烟碱依赖症状——烟碱依赖的风险和严重程度受ENDS产品特性（烟碱浓度、风味、装置类型和品牌）的影响，

---

1 仅提供 NASEM 认为具有确凿、实质性和适度证据的结论，而不是证据有限、不充分或没有可用证据的结论。确凿、实质性和适度的科学证据分别允许得出确定的结论、具有轻微局限性的结论和具有局限性的一般结论。证据强度指的是关联的确定性，但不一定指关联的程度。使用可靠的文献综述（例如 NASEM 的文献综述）的优势在于，其结论是基于综述时对综合证据的系统性和方法性概述。然而，对 EN & NNDS 的研究正在迅速发展，这意味着一些新的研究可能与系统性综述的结论相矛盾。本文不讨论尚未纳入可靠系统综述的新研究结果，除非它们提供了压倒性和无可争议的证据。

但是ENDS的依赖风险和严重程度似乎比燃烧型卷烟低；
- EN＆NNDS气溶胶可能导致某些人体细胞功能失常——目前尚不清楚这对长期使用EN＆NNDS的后果意味着什么，但这可能会增加某些疾病的风险，例如心血管疾病、癌症和不良生殖影响，尽管风险可能低于燃烧型卷烟烟气；
- 从经常使用燃烧型卷烟完全转换为EN＆NNDS可以减少多个器官系统的短期不良健康后果。

有适度证据表明：
- EN＆NNDS的使用会增加青少年的咳嗽和气喘，并且与哮喘恶化的增加有相关性；
- 使用EN＆NNDS对健康的积极和消极影响适用于这些产品的使用是在没有同时抽吸烟草制品的情况下，但是EN＆NNDS使用者中有很大一部分同时抽吸烟草制品（被称为双重或多重使用者）。

问题是——继续抽吸卷烟的EN＆NNDS使用者是否会降低健康风险？NASEM审查得出的结论是，与仅抽吸燃烧型卷烟的吸烟者相比，吸烟者长期使用电子烟（双重使用）是否会改变发病率或死亡率，目前尚无可用的证据。然而，最近的证据表明，双重使用者的氧化应激水平高于吸烟者[17]，并且吸烟者同时使用EN＆NNDS可能会导致心肺健康风险，特别是涉及呼吸系统的风险[18]。

关于美国使用EN＆NNDS引起的肺部疾病案例的说明

在起草本简报期间，美国疾病控制与预防中心（CDC）报告了与使用电子烟和雾化产品有关的肺部疾病暴发[19]。截至2020年1月7日，美国50个州向CDC报告的病例超过2500例。27个州已确认有近60人死亡。

美国CDC已将维生素E醋酸酯确定为电子烟或雾化产品使用相关肺损伤（EVALI）患者中一种值得关注的化学物质。CDC实验室检测了从10个州提交的29名EVALI患者肺部收集的液体样本，发现所有样本中都含有维生素E醋酸酯。维生素E醋酸酯用作添加剂，尤其是用作含四氢大麻酚的电子烟或雾化产品的增稠剂。

另一项研究得出的结论是，与仅吸烟者相比，双重使用者由于继续吸烟并没有降低有害毒物的暴露量[20]。一种可能的解释是，双重使用者包括

多种烟草和EN&NNDS使用行为，每种行为都有不同的动机[21]。双重使用可能不仅代表减少吸烟或戒烟的过渡阶段，还可能包括仍然依赖吸烟来解决其对EN&NNDS体验的不满、规避无烟政策或只是为了遵守社会群体规范以及应对与吸烟有关的耻辱感的EN&NNDS使用者[22]。

电子烟碱/非烟碱传输系统气溶胶的二手烟暴露

EN&NNDS使用者直接从设备吸入气溶胶，并将部分气溶胶呼出到空气中，非吸烟者可能会吸入这些气溶胶。因此，与背景水平相比，使用EN&NNDS会增加室内环境中空气传播的颗粒物和烟碱的浓度[16]。一些研究表明，在使用EN&NNDS的过程中，一些挥发性有机化合物也被呼出到环境中。空气中这些物质的浓度随密闭空间中使用者数量的增加而增加。与燃烧型卷烟相比，来自EN&NNDS气溶胶的二手烟烟碱和颗粒物的暴露量较低[16]，但高于世界卫生组织《烟草控制框架公约》（WHO FCTC）建议的无烟水平[23]。

呼出气溶胶暴露的健康影响

目前还没有研究评估二手EN&NNDS暴露对健康的影响，因此，呼出气溶胶暴露对健康的风险仍然未知。但是，预计会给非吸烟者带来一些健康风险，尽管风险低于二手烟草烟气暴露。

电子烟碱/非烟碱传输系统对戒烟和起始吸烟的作用

电子烟碱/非烟碱传输系统对成年人戒烟的作用

NASEM审查得出的结论是，随机对照试验的证据不足以证明ENDS作为戒烟辅助工具与不治疗或经批准的戒烟治疗相比的有效性[16]，但该审查未包括最近的一项试验，其结果与该结论相反[24]。但是，有适度证据表明，某些吸烟者可以通过频繁或密集使用某些类型的ENDS来成功戒烟[16]，而其他吸烟者的戒烟效果则没有差别，甚至无法戒烟[25]。

电子烟碱/非烟碱传输系统对青少年起始吸烟的作用

有适度的证据表明，尝试使用EN&NNDS的非吸烟青少年此后尝试吸烟的可能性至少高出一倍[16]。然而，目前可用的数据并不能证明这种明显的关联是因果关系。尽管一些作者认为使用ENDS和吸烟是由于对危险行为的共同潜在倾向而独立开始的，但另一些作者认为，在社交学习框架中，使用ENDS和吸烟之间的相似性促进了从一种产品向另一种产品的轨迹。

调味剂对电子烟碱/非烟碱传输系统使用的作用

市场上用于EN&NNDS的电子烟烟液有15 000多种独特口味[26,27]。调味剂分为两大类：烟草调味剂和具有强烈非烟草气味或味道的调味剂。后者被认为是所谓的特征风味，其主要类别包括薄荷醇/薄荷、坚果、香辛料、咖啡/茶、酒精、其他饮料、水果、糖果和其他甜味剂[28]。

风味是EN&NNDS最具吸引力的特征之一，被描述为青少年使用ENDS的主要动机。它们可以改变人们对EN&NNDS的期望和从中获得的奖励，包括烟碱效应[29,30]。在电子烟烟液容器和供应商网站上的广告通常包含传达产品感官吸引力的图像和口味说明[31]。

风味似乎在促进从燃烧型烟草制品转向EN＆NNDS中发挥了一定作用[32-34]，在增加青少年对EN＆NNDS的使用率方面发挥了重要作用[35~37]，比成年人更显著[38]。与年长者相比，年轻人和青壮年风味电子烟烟液的使用率通常较高。与传统吸烟者相比，非吸烟者的使用率也更高[39]。对风味烟碱产品的偏好和需求似乎适用于传统卷烟和EN＆NNDS。使用者倾向于从所有可用烟碱产品风味中寻求奖励[40]。换句话说，当期望的烟碱产品不能提供想要的风味时，一部分使用者可能会从第二选择的烟碱产品中寻求奖励。

关键信息与结论

EN&NNDS并非无害。尽管尚未充分研究其对发病率和死亡率的长期影响，但EN&NNDS对于青少年、孕妇和从未吸烟的成年人并不安全。虽然预计这些人群使用EN&NNDS可能会增加其健康风险，但成年吸烟者（孕妇除外）完全从燃烧型卷烟转向使用未掺杂且经过适当监管的EN&NNDS，则可能会降低其健康风险。WHO[41]、NASEM[16]和CDC[42]都已经意识到了这种可能性。

正如WHO[41]指出的那样，对EN&NNDS采取任何政策的关键是"适当地监管这些产品，以最大限度地减少可能导致烟草流行的后果，并优化对公共卫生的潜在益处"，以及"避免非吸烟者尤其是青少年开始使用烟碱，同时最大限度地提高吸烟者的潜在益处"。鉴于现有的科学证据，以及并非所有国家都具备所需的监管和监测能力，实现这种监管平衡具有挑战性[43]。决定对EN&NNDS进行监管的WHO成员国可以考虑以下措施，以实现WHO FCTC缔约方会议（COP）设定的政策目标[44]：

- 防止非吸烟者和青少年开始使用EN&NNDS，特别注意弱势群体。
- 尽量降低EN&NNDS使用者的潜在健康风险，并保护非使用者免于暴露于其释放物。
- 防止对EN&NNDS做出未经证实的健康声明，并保护烟草控制活动不受与EN&NNDS相关的所有商业利益和其他既得利益的影响，包括烟草行业的利益。

决定监管EN&NNDS的国家应考虑：

- 注意任何监管措施在将市场转向任何特定类型的EN&NNDS产品方面的意外后果。
- 监管以医疗产品和治疗设备做出健康声明的EN&NNDS，并在科学验证此类声明后授权其销售。
- 禁止或限制EN&NNDS的广告、促销和赞助，规范销售渠道（包括在线销售），并严格执行有关最低购买年龄的法律，同时认识到限制未成年人和成年人获得烟草制品，使其在使用EN&NNDS时难以转向使用卷烟是至关重要的。

通过标准化以下方面，最大限度地降低EN&NNDS使用者的健康风险；

- 根据有效的电气设备安全法规，包括电气和电子设备的废物和安全处置法规，规范设备和EN&NNDS组件的生产；
- 电子烟烟液的成分，限制每个烟弹或每瓶中可用的烟碱量，并避免使用某些成分，例如致癌物、诱变剂或生殖毒素，促进吸入或烟碱摄入的成分以及添加剂，例如氨基酸、咖啡因、着色剂、必需脂肪酸、葡萄糖醛酸内酯、益生菌、牛磺酸、维生素和矿物质营养素——现有证据不足以建议禁止（或不禁止）的某些可能对儿童有吸引力的口味；
- 电子烟烟液的包装要求使用具有儿童防护功能的容器，并在ENDS上标识以告知使用者该产品具有成瘾性。
- 在证明二手气溶胶不会对非吸烟者构成健康风险之前，禁止在所有室内场所或禁止吸烟的场所使用EN&NNDS来最大限度地降低非使用者的健康风险。
- 限制EN&NNDS中允许的特定调味剂的含量和种类，以减少青少年开始使用EN&NNDS。
- 建立监测系统以监测EN&NNDS消费模式的演变并调查涉及EN&NNDS的健康或安全事件——鉴于目前对市场动态的了解，对于各国而言，开始监测市场中的EN&NNDS产品并评估监管对价格和消费的影响极为重要（包括按使用强度、设备类型、电子烟烟液的成分和使用原因，以及群体统计特征和吸烟状况来监测EN&NNDS的人群使用模式）；随着市场的快速发展，可能需要随着时间推移对税收方式进行调整。

此外，决定征收消费税的国家应考虑：

- 采用由国家税收管理水平、产品监管和烟草控制政策所决定的最佳税收结构。例如，税收管理和产品监管较强的国家可能会发现，选择特定消费税更有利，而税收管理强但产品监管弱的国家可能会发现从价税制度是更好的一种选择。
- 无论环境如何，设置产品特性以提高税收结构的有效性。
- 以与该国烟草制品相同的方式征税（在大多数国家/地区，征收是在源头，即生产和进口环节进行的）。

某些类型的ENDS可在某些情况下帮助某些吸烟者戒烟，但目前还没有足够的证据来全面建议所有吸烟者使用任何类型的EN&NNDS作为戒烟辅助工具。

  关于EN&NNDS政策（无论其性质如何）的最后一个重要的注意事项是，如果同时实施强有力的烟草控制政策，以遏制从EN&NNDS使用到吸烟的任何潜在轨迹，那么这种政策将受益匪浅。

参 考 文 献[1]

[1] Dai H, Leventhal A. Prevalence of e-cigarette use among adults in the United States, 2014–2018. JAMA 2019;322(18):1824–27. doi:10.1001/jama.2019.15331.

[2] Special Eurobarometer 458: attitudes of Europeans towards tobacco and electronic cigarettes. Brussels: European Commission, Directorate General for Health and Food Safety; 2017 (https://ec.europa.eu/commfrontoffice/publicopinion/index.cfm/ResultDoc/download/DocumentKy/79003).

[3] Vaping in England: evidence update summary February 2019. London: Public Health England; 2019 (https://www.gov.uk/government/publications/vaping-in-england-an-evidence-updatefebruary-2019/vaping-in-england-evidence-update-summary-february-2019#vaping-in-youngpeople).

[4] New Zealand Health Survey: use e-cigarettes once a month. In: Annual update of key results 2017/18. New Zealand Health Survey [website]: Wellington: Ministry of Health; 2019 (https://minhealthnz.shinyapps.io/nz-health-survey-2018-19-annual-data-explorer/_w_01f170d8/#!/explore-indicators).

[5] Appendix XI, table 11.2 – adult tobacco survey smokeless tobacco or e-cigarettes. In: WHO report on the global tobacco epidemic 2019 [website]. Geneva: World Health Organization; 2019 (https://www.who.int/tobacco/global_report/en/).

[6] Special Eurobarometer 429: attitudes of Europeans towards tobacco and electronic cigarettes. Brussels: European Commission, Directorate General for Health and Food Safety; 2015 (https://ec.europa.eu/commfrontoffice/publicopinion/archives/ebs/ebs_429_en.pdf).

[7] Prevalence of e-cigarette use. In: Reid JL, Hammond D, Tariq U, Burkhalter R, Rynard VL, Douglas O. Tobacco use in Canada: patterns and trends, 2019 edition. Waterloo (ON): Propel Centre for Population Health Impact, University of Waterloo; 2019:90–7(https://uwaterloo.ca/tobacco-usecanada/sites/ca.tobacco-use-canada/files/uploads/files/tobacco_use_in_canada_2019.pdf).

[8] Appendix XI, table 11.4 – youth tobacco surveys smokeless tobacco or e-cigarettes. In: WHO report on the global tobacco epidemic 2019 [website]. Geneva: World Health Organization; 2019 (https://www.who.int/tobacco/global_report/en/).

[9] Yoong SL, Stockings E, Chai LK, Tzelepis F, Wiggers, Oldmeadow C et al. Prevalence of electronic nicotine delivery systems (ENDS) use among youth globally: a systematic review and metaanalysis of country level data. Aust NZ J Public Health 2018;42(3):303–8. doi:10.1111/1753-6405.12727.

[10] Cullen K, Ambrose B, Gentzke A, Apelberg B, Jamal A, King B. Notes from the field: use

---

1 所有链接均于 2020 年 1 月 13 日访问。

of electronic cigarettes and any tobacco product among middle and high school students — United States, 2011–2018. MMWR Morb Mortal Wkly Rep. 2018;67(45):1276–77. doi:10.15585/mmwr.mm6745a5.

[11] Use of e-cigarettes among young people in Great Britain. London: Action on Smoking and Health; 2019 (https://ash.org.uk/wp-content/uploads/2019/06/ASH-Factsheet-Youth-Ecigarette-Use-2019.pdf).

[12] E-cigarette use. In: Reid JL, Hammond D, Tariq U, Burkhalter R, Rynard VL, Douglas O. Tobacco use in Canada: patterns and trends, 2019 edition. Waterloo (ON): Propel Centre for Population Health Impact, University of Waterloo; 2019:89–105 (https://uwaterloo.ca/tobacco-use-canada/sites/ca.tobacco-use-canada/files/uploads/files/tobacco_use_in_canada_2019.pdf).

[13] Hammond D, Reid JL, Rynard V, Fong GT, Cummings KM, McNeill A et al. Prevalence of vaping and smoking among adolescents in Canada, England, and the United States: repeat national cross-sectional surveys. BMJ 2019;365:l2219. doi:10.1136/bmj.l2219.

[14] Historical NYTS data and documentation. In: Centers for Disease Control and Prevention [website]. Atlanta (GA): Centers for Disease Control and Prevention; 2019 (https://www.cdc.gov/tobacco/data_statistics/surveys/nyts/data/index.html).

[15] McNeill A, Brose LS, Calder R, Bauld L, Robson D. Evidence review of e-cigarettes and heated tobacco products 2018. A report commissioned by Public Health England. London: Public Health England; 2018 (https://assets.publishing.service.gov.uk/government/uploads/system/uploads/attachment_data/file/684963/Evidence_review_of_e-cigarettes_and_heated_tobacco_products_2018.pdf).

[16] The National Academies of Sciences, Engineering, Medicine. Public health consequences of e-cigarettes. Washington (DC): The National Academies Press; 2018 (https://www.ncbi.nlm.nih.gov/pubmed/29894118).

[17] POS5-51: PATH study wave 1 biomarkers of inflammation and oxidative stress among adult e-cigarette and cigarette users [research poster]. In: SNRT 25 Rapid Response Abstracts. San Francisco (CA): Society for Research on Nicotine and Tobacco; 2019:24 (https://cdn.ymaws.com/www.srnt.org/resource/resmgr/SRNT19_Rapid_Abstracts.pdf).

[18] Wang J, Olgin J, Nah G, Vittinghof E, Cataldo JK, Pletcher MJ et al. Cigarette and e-cigarette dual use and risk of cardiopulmonary symptoms in the Health eHeart Study. PLoS ONE 2018;13(7):e0198681. doi:10.1371/journal.pone.0198681.

[19] Outbreak of lung injury associated with e-cigarette use, or vaping, products. In: Centers for Disease Control and Prevention [website]. Atlanta (GA): Centers for Disease Control and Prevention; 2019 (https://www.cdc.gov/tobacco/basic_information/e-cigarettes/severe-lungdisease.html#latest-outbreak-information).

[20] Goniewicz ML, Smith DM, Edwards KC, Blount BC, Caldwell KL, Feng J et al. Comparison of nicotine and toxicant exposure in users of electronic cigarettes and combustible cigarettes. JAMA Netw Open 2018;1(8):e185937. doi:10.1001/jamanetworkopen.2018.5937.

[21] Borland R, Murray K, Gravely S, Fong GT, Thompson ME, McNeill A et al. A new

classification system for describing concurrent use of nicotine vaping products alongside cigarettes (socalled "dual use"): findings from the ITC-4 Country Smoking and Vaping Wave 1 Survey. Addiction 2019;114(S1):24–34. doi:10.1111/add.14570.

[22] Robertson L, Hoek J, Blank M, Richards R, Ling P, Popova L. Dual use of electronic nicotine delivery systems (ENDS) and smoked tobacco: a qualitative analysis. Tob Control 2019;28:13–9. doi: 10.1136/tobaccocontrol-2017-054070.

[23] Guidelines for implementation of Article 8: protection from exposure to tobacco smoke. WHO Framework Convention on Tobacco Control. Geneva: World Health Organization; 2007 (https://www.who.int/fctc/guidelines/adopted/article_8/en/).

[24] Hajek P, Phillips-Waller A, Przulj D, Pesola F, Myers Smith K, Bisal N et al. A randomized trial of e-cigarettes versus nicotine-replacement therapy. New Engl J Med. 2019;380(7):629–37. doi:10.1056/nejmoa1808779.

[25] Peruga A, Eissenberg T. Clinical pharmacology of nicotine in electronic nicotine delivery systems. In: WHO TobReg: report on the scientific basis of tobacco product regulation. Seventh report of a WHO study group. Geneva: World Health Organization; 2019:31–74 (WHO Technical Report Series No. 1015; https://apps.who.int/iris/bitstream/handle/10665/329445/9789241210249-eng.pdf?ua=1).

[26] Zhu SH, Sun JY, Bonnevie E, Cummins SE, Gamst A, Yin L et al. Four hundred and sixty brands of e-cigarettes and counting: implications for product regulation. Tob Control 2014;23(Suppl. 3):iii3–9. doi:10.1136/tobaccocontrol-2014-051670.

[27] Hsu G, Sun J, Zhu S. Evolution of electronic cigarette brands from 2013–2014 to 2016–2017: analysis of brand websites. J Med Internet Res. 2018;20(3):e80. doi:10.2196/jmir.8550.

[28] Krüsemann E, Boesveldt S, de Graaf K, Talhout R. An e-liquid flavor wheel: a shared vocabulary based on systematically reviewing e-liquid flavor classifications in literature. Nicotine TobRes. 2018;21(10):1310–9. doi:10.1093/ntr/nty101.

[29] Krishnan-Sarin SS, O'Malley S, Green BG, Pierce JB, Jordt SE. The science of flavour in tobacco products. In: WHO study group on tobacco product regulation. Report on the scientific basis of tobacco product regulation. Seventh report of a WHO study group. Geneva: World Health Organization; 2019:125–42 (WHO Technical Report Series No. 1015; https://apps.who.int/iris/bitstream/handle/10665/329445/9789241210249-eng.pdf?ua=1).

[30] Zare S, Nemati M, Zheng Y. A systematic review of consumer preference for e-cigarette attributes: flavor, nicotine strength, and type. PLoS ONE 2018;13(3):e0194145. doi:10.1371/journal.pone.0194145.

[31] Soule EK, Sakuma KK, Palafox S, Pokhrel P, Herzog TA, Thompson N et al. Content analysis of internet marketing strategies used to promote flavored electronic cigarettes. Addict Behav. 2019;91:128–35. doi:10.1016/j.addbeh.2018.11.012.

[32] Farsalinos KE, Romagna G, Tsiapras D, Kyrzopoulos S, Spyrou A, Voudris V. Impact of flavour variability on electronic cigarette use experience: an internet survey. Int J Environ Res Public Health 2013;10(12):7272–82. doi:10.3390/ijerph10127272.

[33] Shiffman S, Sembower MA, Pillitteri JL, Gerlach KK, Gitchell JG. The impact of flavor descriptors on nonsmoking teens' and adult smokers' interest in electronic cigarettes. Nicotine Tob Res. 2015;17(10):1255–62. doi:10.1093/ntr/ntu333.

[34] Tackett AP, Lechner WV, Meier E, Grant DM, Driskill LM, Tahirkheli NN et al. Biochemically verified smoking cessation and vaping beliefs among vape store customers. Addiction 2015;110(5):868–74. doi:10.1111/add.12878.

[35] Audrain-McGovern J, Strasser AA, Wileyto EP. The impact of flavoring on the rewarding and reinforcing value of e-cigarettes with nicotine among young adult smokers. Drug Alcohol Depend. 2016;166:263–7. doi:10.1016/j.drugalcdep.2016.06.030.

[36] Kong G, Morean ME, Cavallo DA, Camenga DR, Krishnan-Sarin S. Reasons for electronic cigarette experimentation and discontinuation among adolescents and young adults. Nicotine Tob Res. 2015;17(7):847–54. doi:10.1093/ntr/ntu257.

[37] Krishnan-Sarin S, Morean ME, Camenga DR, Cavallo DA, Kong G. E-cigarette use among high school and middle school adolescents in Connecticut. Nicotine Tobacco Res. 2015;17(7):810–8. doi:10.1093/ntr/ntu243.

[38] Morean ME, Butler ER, Bold KW, Kong G, Camenga DR, Dana A et al. Preferring more e-cigarette flavors is associated with e-cigarette use frequency among adolescents but not adults. PLoS ONE 2018;13(1):e0189015. doi:10.1371/journal.pone.0189015.

[39] Goldenson NI, Leventhal AM, Simpson KA, Barrington-Trimis JL. A review of the use and appeal of flavored electronic cigarettes. Curr Addict Rep. 2019;6(2):98–113. doi:10.1007/s40429-019-00244-4.

[40] Buckell J, Marti J, Sindelar JL. Should flavours be banned in cigarettes and e-cigarettes? Evidence on adult smokers and recent quitters from a discrete choice experiment. Tob Control. 2019;28(2):168–75. doi:10.1136/tobaccocontrol-2017-054165.

[41] Provisional agenda item 5.5.2: electronic nicotine delivery systems and electronic non-nicotine delivery systems (ENDS/ENNDS). Report by WHO. In: Conference of the Parties to the WHO Framework Convention on Tobacco Control: seventh session, Delhi, India, 7–12 November 2016. Geneva: World Health Organization; 2016 (Document FCTC/COP/7/11; https://www.who.int/tobacco/communications/statements/eletronic-cigarettes-january-2017/en/).

[42] Electronic cigarettes: what's the bottom line? Atlanta (GA): Centers for Disease Control and Prevention; 2019 (https://www.cdc.gov/tobacco/basic_information/e-cigarettes/pdfs/Electronic-Cigarettes-Infographic-508.pdf).

[43] Tobacco product regulation: basic handbook. Geneva: World Health Organization; 2018 (https://apps.who.int/iris/handle/10665/274262).

[44] Decision: electronic nicotine delivery systems and electronic non-nicotine delivery systems. In: Conference of the Parties to the WHO Framework Convention on Tobacco Control: sixth session, Moscow, Russian Federation,13–18 October 2014. Geneva: World Health Organization; 2014 (document FCTC/COP6(9); https://apps.who.int/gb/fctc/E/E_cop6.htm).

电子烟碱/非烟碱传输系统监管国家案例

摘　　要

电子烟碱/非烟碱传输系统（EN&NNDS）是一类使用电加热线圈将电子烟烟液转化为气溶胶供使用者吸入的异质性产品。在全球范围内，各国政府采用不同的方法对市场上的EN&NNDS进行监管。本简报举例说明了三种典型的监管方法，包括禁止EN&NNDS的销售，对这些产品应用烟草控制法规，以及就销售、营销、包装、产品监管、报告/通告、税收以及在工作场所和公共场所的使用制定一系列详细的具体法规和建议。

关键词：电子烟碱传输系统（ENDS）；电子非烟碱传输系统（ENNDS）；监管；案例研究

致　　谢

这份简报由世界卫生组织欧洲地区办事处顾问Armando Peruga撰写，世界卫生组织总部科学家Sarah Galbraith-Emami，技术官员Marine Perraudin，项目经理Kristina Mauer-Stender及世界卫生组织欧洲地区办事处非传染性疾病司和生命历程健康促进司烟草控制计划顾问Elizaveta Lebedeva参与编写。

作者还要感谢世界卫生组织欧洲地区办事处非传染性疾病和生命历程健康促进司代理司长Nino Berdzuli对本简报的编写工作给予总体指导和支持。

本项目由俄罗斯联邦政府和德国政府提供资金。

引 言

电子烟碱传输系统（ENDS）是一类混合产品，使用电加热线圈将电子烟烟液转化为气溶胶，供使用者吸入。电子烟烟液通常由丙二醇、甘油或其混合物制成，均含有烟碱，并且可能含有调味剂。当电子烟烟液中不含烟碱时，被称为电子非烟碱传输系统（ENNDS）。ENDS和ENNDS统称为EN&NNDS。

各国政府采用不同的方法来监管其市场中的EN&NNDS。以巴西、加拿大、韩国和英国四个国家为例来说明这些不同的监管方法。

加拿大和英国允许销售EN&NNDS，并以适用于这些产品的具体规范对其进行监管，但两国都有自己的特点。加拿大实行联邦政府制度，国家（或联邦）一级政府通过其行政、立法或司法部门对公共生活领域的某些领域进行监管。国家以下各级政府可以补充与卫生相关的联邦法规，也可以自行监管。在英国，政府明确提倡吸烟者从传统烟草转向使用ENDS来戒烟。韩国也允许销售EN&NNDS，并将传统烟草制品的现行法规应用于ENDS，但对ENNDS不适用。巴西实际上禁止销售ENDS和ENNDS。

四个国家中有三个国家拥有强有力的控烟环境，表现为在六项MPOWER措施[1]中至少有四项的实施达到最高水平[1]。下文将介绍每个国家EN&NNDS的流行率以及对EN&NNDS的一般监管方法。附录A和附录B按政策领域列出了国家或联邦法规，对加拿大则是介绍了各省辖区的具体法规。

图3-1比较了巴西、韩国、加拿大、英国四个国家的吸烟者和戒烟者中每天使用ENDS的比例。

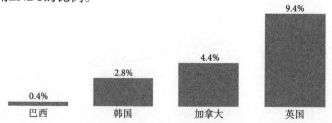

图3-1　2016年四个国家18岁以上吸烟者和最近戒烟者混合人群中每天使用ENDS的比例[2]

1　MPOWER的六项措施是：监测烟草使用和预防政策，保护人们免受烟草烟气的危害，提供戒烟帮助，警示烟草危害（健康警示），执行烟草广告、促销和赞助禁令，提高烟草税。

巴　西

巴西EN＆NNDS使用情况的总结请参见专栏3-1[2]。

> **专栏3-1　EN＆NNDS在巴西的使用情况**
>
> 　　巴西的最新数据来自2016年，且仅涵盖里约热内卢、圣保罗和阿雷格里港的1340名成年吸烟者和近期戒烟者样本中ENDS的流行率。有1%的受访者每月至少使用一次ENDS，0.4%的受访者每天至少使用一次。
>
> 　　资料来源：Gravely[2]

　　实际上，巴西禁止EN＆NNDS。巴西国家卫生监督局（ANVISA）合议理事会（RDC）第46/2009号决议[3]禁止电子烟设备的营销、进口和广告。电子烟设备包括EN＆NNDS和加热型烟草制品（HTP），但是，希望在巴西销售EN＆NNDS的制造商可以提交产品注册申请，前提是具备必要的背景资料证明设备的功效、有效性和安全性。在这种情况下，ANVISA会审查提供的信息并决定是否对产品进行注册，从而允许其上市销售。迄今为止，尚未收到EN＆NNDS注册申请。如果已注册，则有关烟草控制的法律将适用于EN＆NNDS，并且将禁止在所有室内公共场所、工作场所和公共交通工具中使用。

　　ANVISA对EN＆NNDS的政策仍然是禁用。当时，该机构的决定是基于缺乏有关这些产品声明的科学数据，但ANVISA通过审查有关EN＆NNDS潜在健康风险和益处的科学证据（2016年）[4]、技术小组（2018年）[5]和有行业和公共卫生组织参与的公开听证会（2019年）[6]来定期考虑其立场（听证会所有发言请参见ANVISA的链接[7]）。

加 拿 大

加拿大EN&NNDS使用情况的总结请参见专栏3-2[8]。

专栏3-2　EN&NNDS在加拿大的使用情况

2017年：

- 15岁以上的加拿大人中，过去30天使用和每天使用EN&NNDS的比例分别为2.9%和1.0%；
- 12.2%的当前吸烟者和2.4%的非吸烟者在过去30天内使用过EN&NNDS；
- 约有64%的成年EN&NNDS使用者自述上一次使用电子烟时，电子烟烟液中含有烟碱，尽管当时含烟碱的电子烟烟液尚未批准在加拿大销售；
- 水果味和烟草味是最近使用的EN&NNDS中最常被提及的风味。

2016~2017年间，7~9年级的青少年中：

- 5.4%在过去30天内使用过EN&NNDS产品；
- 三分之二的当前吸烟者过去30天内使用过EN&NNDS，而非吸烟者中这一比例为5%。

资料来源：Reid等[8]

加拿大联邦法律《烟草和电子烟产品法》（TVPA）[9]在2018年才允许销售ENDS，但ENNDS的销售一直是允许的。在撰写本简报时，TVPA的ENDS法规尚未完全在联邦一级制定。但是，加拿大卫生部就此类规定进行了公众咨询[10]；总结了收到的意见[11]，政府于2019年12月发布了拟议的法规。这些法规将征询公众意见到2020年1月底[12]。如果获得批准，该法规将采取更多的控制措施：①进一步限制电子烟产品的促销，包括在销售点

和在线销售；②要求在广告上标注健康警示；③禁止生产具有某些风味或风味成分的电子烟产品；④限制电子烟产品中烟碱的浓度和输送。

当前的联邦法律将做出健康声明与不做出健康声明的EN&NNDS分开考虑。如果做出健康声明，除了TVPA之外，还在联邦一级受《食品和药品法》[13]及其法规的监管。如果EN&NNDS没有做出健康声明，则受2018年修订的《加拿大消费品安全法》[14]和2018年的TVPA的监管。

TVPA旨在"防止使用电子烟产品导致年轻人和非烟草制品使用者使用烟草制品"[9]。它规定了全国范围内电子烟产品的最低销售年龄，并严格限制其促销活动，包括禁止对年轻人有吸引力的生活方式广告或促销活动。

所有电子烟液还应遵守2001年《消费化学品和容器法规》[15]。《非吸烟者健康法》规定了在联邦工作场所允许使用和不允许使用EN&NNDS的场所[16]。各省有权扩展法律的适用范围，包括在不属于联邦管辖的场所使用和不使用EN&NNDS的情况。

根据加拿大税法，EN&NNDS不被视为烟草制品。目前仅适用常规的一般销售税，其中包括省级销售税部分[17]。然而，有两个省在2019年宣布2020年提高EN&NNDS税率。不列颠哥伦比亚省政府把EN&NNDS的省级销售税从7%提高到20%[18]。该税将适用于所有EN&NNDS设备、设备中使用的任何物质以及零部件和配件。不列颠哥伦比亚省的新法规还将要求电子烟产品使用素包装和健康警示，并限制在年轻人经常光顾的区域做公共广告。艾伯塔省政府推出财政计划[19]，其中包括2020年对EN&NNDS产品零售价征收20%的税。该税将适用于所有电子烟烟液，包括单独出售的大麻电子烟烟液和自制电子烟产品，例如丙二醇、植物甘油、烟碱溶液和调味剂，以及所有电子烟设备和相关配件[20]。在这两个省中，提税的理由是为了遏制年轻人使用EN&NNDS的增加。

加拿大卫生部积极监测个人和企业遵守TVPA的情况。在2019年检查了3000多家零售店，同时跟踪EN&NNDS的在线销售和推广情况，并在必要时采取措施。加拿大卫生部还通过广告、社交媒体和学校体验活动开展公众教育活动，以提高13~18岁的加拿大青少年对使用电子烟产品相关危害和风险的认识[21,22]。

附录A包含适用于电子烟产品的联邦法规的说明。加拿大是一个联邦制国家，其部分地方辖区（各省）已经在其权限范围内监管EN&NNDS。在12个省和地区中，有8个制定了EN&NNDS有关法律（有关规定的摘要，请参见附录B）。

韩　国

韩国EN&NNDS使用情况的总结请参见专栏3-3[2,23,24]。

专栏3-3　EN&NNDS在韩国的使用情况

2017年，韩国国民健康与营养调查估计，在19岁以上的成年人中，每月至少使用一次EN&NNDS的比例为2.3%。根据2018年韩国青少年风险行为网络调查，在13~18岁青少年中，这一比例为2.7%。

对成年吸烟者的一项研究表明，有5.5%的人每月至少使用一次EN&NNDS，2.8%的人每天使用。

资料来源：Gravely[2]；WHO[23];WHO[24]

2007年，政府修订了《烟草商业法》[25]的适用范围，使其不仅适用于使用烟叶制造的产品，还适用于不使用烟叶作为吸入原料制造的产品，例如ENDS，但《药品事务法》[26]所涵盖的作为药品或非药品使用的产品除外。2017年《国民健康促进法》[28]实施令[27]第27-2条将ENDS明确命名为电子烟，并将其定义为"使用电子设备将含烟碱的溶液或烟丝通过呼吸器官吸入人体而产生与吸烟相同效果的产品"。

ENDS和HTP在韩国被归类为"电子烟草"产品，因此大多数烟草控制法规都适用于ENDS和HTP。

禁止吸烟的场所禁止使用电子烟。除了大学以外，室内医疗保健和教育机构的所有区域都禁止吸烟。在工作场所和公共场所的指定非吸烟区也禁止吸烟。

在电视、广播、广告牌以及其他户外设施上宣传ENDS是非法的。促销折扣等一些营销形式也被禁止。但是，执法工作一直面临挑战，典型的例子是英美烟草（韩国）公司拒绝撤下一个音乐录影带形式的广告，理由是它展示的是"电子设备"，而不是装有实际烟碱产品的容器。同时，英美烟草公司为其电子烟设备提供了声势浩大的50%折扣[29]。其他事例还包括一家ENDS公司通过在年轻人的在线论坛中宣传电影并提供赢取免费电影票的机会来瞄准年轻人；在青少年电影中植入EN&NNDS；以及赞助社区青年交响乐团[30]。

ENDS包装上必须带有占包装主表面的50%的图形健康警示[31]。ENDS包装和广告中应包含健康警示文字，说明其中含有有害物质，例如烟草特有亚硝胺和甲醛。

ENDS的烟草消费税为每毫升烟碱溶液628韩元（约合0.5美元）[32]。除了每20个烟弹收取24韩元（约0.02美元）的废弃物处理税和10%的增值税（VAT）外，ENDS还应缴纳其他税费（国民健康促进税、地方教育税和个人消费税）。总的来说，ENDS的税率为每毫升烟碱溶液1799韩元（约合1.5美元）[33]。

2019年，美国使用掺杂的EN&NNDS电子烟液而导致肺病流行，这促使韩国政府采取具体行动，遏制年轻人使用这些产品。政府在四个方面采取行动：修订立法、执行现行法律、监督和教育。

卫生和福利部已向国民议会提交了对法律的修改意见，以弥补烟草类别定义中的某些漏洞，将新型产品纳入其中，并禁止某些风味。尚不清楚这些宣布的法律修改的具体内容。通过这项立法，政府还将要求电子烟制造商提交有关其产品成分和添加剂的更详细的信息。

在执行现有法律时，政府的重点是减少电子烟的非法销售，包括在线销售和对未成年人销售。同时，出于安全原因，韩国技术标准局正在阻止非法电池的分销和销售，全国各地的政府部门都在进行检查，以执行在公共场所无烟法规，并特别关注ENDS和HTP的使用[34]。

政府正在通过现有的消费者风险监控系统密切监测可能与使用EN&NNDS有关的潜在肺部疾病病例。卫生和福利部正在开展健康教育运动，旨在提供与EN&NNDS相关的健康风险信息，并为吸烟者提供戒烟咨询。

英　国

英国EN＆NNDS使用情况的总结请参见专栏3-4[35,36]。

专栏3-4　EN＆NNDS在英国的使用情况

根据"欧洲晴雨表"报告，2017年，5.6%的成年人每月至少使用一次EN＆NNDS，约有300万使用者。在欧盟国家中，英国是EN＆NNDS当前使用率最高的国家。2017年，18%~20%的当前吸烟者也是EN＆NNDS使用者，而大约9%的戒烟者和0.3%~0.6%的从不吸烟者是EN＆NNDS使用者。

2017年，尝试过两次以上EN＆NNDS的成年人中，有42%表示他们这样做是为了帮助完全戒烟。根据调查，2015~2017年间，在11~16岁青少年中，至少每周使用EN＆NNDS的比例为1%~3%，该年龄段从未吸烟者中经常使用EN＆NNDS的比例为0.1%~0.5%。

资料来源：下议院科学技术委员会[35]；欧洲委员会[36]

截至2019年12月，英国仍是欧盟（EU）的一部分，因此受《欧洲烟草制品指令（TPD）》[37]的约束，该指令于2016年5月生效，并通过2016年《烟草及相关产品法规》纳入英国法律[38]。

TPD并不管制具有医疗用途的ENDS。作为消除烟碱成瘾或以显著方式恢复、纠正或改变生理功能或用于医疗目的的补救措施，或以其他方式用于医疗目的的ENDS和补充容器，须遵守2001年11月6日欧洲议会和理事会关于人用医药产品共同体法规的第2001/83/EC号指令，1993年6月14日理事会关于医疗器械的法规第93/42/EEC号指令，以及2004年3月31日欧洲议会和理事会第726/2004/EC/号法规[39]的约束。

在英国，根据这些转化的指令和法规[40-42]，ENDS属于药品和保健品管理局（MHRA）的职权范围。

TPD不涵盖ENNDS，欧盟成员国可以自行对其监管。ENNDS在英国受《通用产品安全条例2005》[43]的监管。

英国对无医疗用途的ENDS的监管：

- 规定电子烟液容器的安全标准，例如儿童防护和防篡改，防止破裂和泄漏，并具有确保重新加注时不会泄漏的装置；
- 规定一次性ENDS、一次性烟弹和烟罐中，贮液容器的体积小于或等于（≤）2 mL；
- 规定专用的填充式电子烟液容器的体积≤10mL，电子烟液中烟碱的最大浓度≤20 mg/mL；
- 禁止在电子烟液中使用以下物质：①使人产生烟草制品对健康有益或降低健康风险印象的添加剂，例如维生素；②与能量和活力相关的添加剂，例如咖啡因或牛磺酸；③对释放物有着色特性的添加剂；④在加热或未加热的情况下对人体健康构成风险的成分（烟碱除外）；
- 要求设备在正常情况下提供一致剂量的烟碱；
- 设置标签要求，例如告知可能的不良影响、致瘾性或毒性的信息，包括成分清单；
- 设置烟碱致瘾性的警示要求；
- 要求所有的ENDS和电子烟液在销售前都必须向MHRA通报；截至2017年底，将近400家生产商提交了关于32407种电子烟液（通告的90%）或设备（通告的10%）的信息；
- 允许消费者和医疗保健专业人员通过电子黄卡报告系统向MHRA报告ENDS和补充容器的副作用和安全问题；
- 禁止向未满18岁的青少年销售和提供EN&NNDS和电子烟烟液；
- 禁止在电视、广播、互联网和特定的印刷出版物上发布ENDS设备和电子烟烟液的广告。

TPD没有规定是否对ENDS征税或如何征税。将这些决定权留给各成员国。EN&NNDS目前在英国作为消费品征税，增值税率为20%。英国没有对EN&NNDS征收特定税。烟草制品的税率要比EN&NNDS高，除了增值税以外，还要征收特定税。在2019年第三季度，10 mL含烟碱的电子烟烟液的价格是20支装万宝路卷烟的4.3倍[44]。

TPD也没有规定可以使用和不能使用ENDS的场所（无ENDS气溶胶区

域），也由成员国自行决定。英国尚未立法限制EN&NNDS的使用场所。威尔士试图限制在某些公共场所使用EN&NNDS，但该法案被否决。但是，仍有许多工作场所、公共场所和运输系统自愿禁止在禁止吸烟的地方使用EN&NNDS。

英国监管工作的特点是积极开展科学和技术辩论。一些报告对讨论产生了影响，特别是英国公共卫生部[1]委托编写并于2019年发布的最新报告[45]，英国皇家内科医学院于2016年发布的报告[46]，英国医学会于2018年发布的报告[47]，以及下议院科学技术委员会2018年的报告[35]。所有这些报告都认为，与吸烟相比，EN&NNDS对健康的危害要小得多，但并非没有风险。它们还指出，缺乏有关长期健康影响的证据。它们表示或暗示ENDS对英国的公众健康有益，因此，将其作为吸烟替代品推广可能会产生巨大的健康收益。同样，它们表明或暗示对EN&NNDS可能的门户效应的担忧在英国并未成为现实[36,48]。

其中几份报告描述了缺乏对ENDS作为戒烟辅助工具的有效性的高质量研究。但是，报告一致认为，大多数研究表明ENDS的使用和戒烟之间存在正相关。英国下议院科学技术委员会的报告建议[35]，国家医疗服务系统（NHS）应为ENDS在精神卫生设施中的作用制定明确的中心政策，将允许患者使用ENDS作为默认政策，除非NHS提出基于证据的不这样做的理由。政府通过一份命令文件对报告做出回应，文件广泛接受了委员会的建议[49]。关于ENDS作为戒烟辅助工具，英国公共卫生协会表示[45]：

将电子烟（一般吸烟者最常用的支持方式）与戒烟服务支持（最有效的支持方式）相结合，应成为所有吸烟者都可选择的推荐方式。

英国国家健康和护理卓越研究所（NICE）[2]于2018年3月建议，卫生和社会服务机构应向吸烟者以及正在使用或有意使用ENDS产品戒烟的人解释，虽然这些产品不是获得许可的药品，但它们是受管制的，许多人发现它们有助于戒烟。NICE还建议使用ENDS的人应完全戒烟，因为任何吸烟行为都是有害的。NHS的长期计划建议为长期接受心理健康和学习障碍服务的吸烟者提供新的通用戒烟方案，其中包括允许吸烟者在住院期间选择改用电子烟。

1　英国公共卫生部是卫生和社会关怀部的一个独立执行机构，负责通过推广更健康的生活方式，向政府提供建议以及支持地方政府、国民健康服务机构和公众的行动，使公众更健康，并缩小不同群体之间的健康差异。

2　NICE是一个独立的政府机构，为改善卫生和社会保健提供国家指导和建议。

下议院科学技术委员会的报告[35]对TPD规定的一些限制表示关注。委员会表示，关于烟罐和补充容器的尺寸、烟碱最大浓度和广告的规范阻碍了ENDS作为戒烟措施的使用，而英国脱离欧盟后这些规范可能会改变。委员会还认为[35]：

吸烟相关产品的税收水平应直接与其健康风险相对应，以鼓励减少有害消费。按照这一逻辑，电子烟的税率应保持最低，而传统卷烟的税率则应保持最高，加热不燃烧产品介于两者之间。

经过协商，广告实践委员会和广播广告实践委员会宣布，将取消在不受TPD监管的媒体中的非广播广告中对ENDS进行健康声明的全面禁令（室外广告、公共交通海报、电影院、传单和直邮）。目前尚不清楚新指南如何在实践中应用。

结　　论

2014年，世界卫生组织《烟草控制框架公约》缔约方大会邀请缔约方[50]：

考虑禁止或管制ENDS/ENNDS，酌情将其作为烟草制品、医药产品、消费品或其他类别，同时考虑到对人类健康的高度保护。

公约秘书处在2018年报告称，有77个缔约方管制或禁止了EN&NNDS[51]。到2019年底，这个数字据报道为98[52]。

本简报举例说明了三种较为典型的监管方法。第一种是禁止销售EN&NNDS。巴西就是一个典型的例子，但现行法律允许政府在制造商提供令人信服的证据来支持变更后重新考虑该决定。第二个是韩国。该国大部分烟草控制法规都适用于EN&NNDS，但有一些例外。最后，加拿大和英国（后者曾是欧盟成员国）制定了一系列详细的具体法规或建议，涉及销售（包括最低年龄）、广告、促销、赞助、包装（儿童防护包装、健康警示标签和商标）、产品规定（烟碱含量/浓度、安全/卫生、成分/风味）、EN/NNDS产品的报告/通告、税收以及在工作场所和公共场所的使用。本简报试图反映两个国家在这一类别中的监管方法，原因有很多，其中包括地方管辖权在EN&NNDS监管中的不同作用，以及加拿大倾向于专门监管ENNDS的某些方面，而英国则没有。

希望这些案例研究能为监管机构和公共卫生倡导者提供信息，帮助他们探究迄今为止在实践中可以使用的EN&NNDS监管方案。

参 考 文 献[1]

[1] WHO report on the global tobacco epidemic, 2019. Geneva: World Health Organization; 2019(https://www.who.int/tobacco/global_report/en/).

[2] Gravely S, Driezen P, Ouimet J, Quah ACK, Cummings KM, Thompson ME et al. Prevalence of awareness, ever-use and current use of nicotine vaping products (NVPs) among adult current smokers and ex-smokers in 14 countries with differing regulations

---

1　所有网络链接均于 2020 年 3 月 18 日访问。

on sales and marketing of NVPs: cross-sectional findings from the ITC Project. Addiction 2019;114(6):1060–73. doi:10.1111/add.14558.

[3] Resolution of the Collegiate Board – RDC number 46, 28 August 2009. Prohibits the sale, import and advertising of any electronic smoking devices, known as electronic cigarettes. Brasília: Ministry of Health, Brazilian Health Regulatory Agency (ANVISA); 2009 (http://portal.anvisa.gov.br/documents/10181/2718376/RDC_46_2009_COMP. pdf/2148a322-03ad-42c3-b5ba-718243bd1919) (in Portuguese).

[4] Electronic cigarettes: what do we know? Study on the composition of vapor and health damage, the role in harm reduction and in the treatment of nicotine dependence. Rio de Janeiro: Ministry of Health, National Cancer Institute José Alencar Gomes da Silva (INCA); 2016 (http://portal.anvisa.gov.br/documents/106510/106594/Livro+Cigarro s+eletr%C3%B4nicos+o+que+sabemos/e8a169d0-fd20-4fdc-b11f-ec9281f49700) (in Portuguese).

[5] Panel debate: electronic smoking devices [news story] In: ANVISA [website]. Brasília: Brazilian Health Regulatory Agency (ANVISA); 2018 (http://portal.anvisa. gov.br/noticias?p_p_id=101_INSTANCE_FXrpx9qY7FbU&p_p_col_id=column-2&p_p_col_pos=1&p_p_col_count=2&_101_INSTANCE_FXrpx9qY7FbU_ groupId=219201&_101_INSTANCE_FXrpx9qY7FbU_urlTitle=painel-debatedispositivos-eletronicos-para-umar&_101_INSTANCE_FXrpx9qY7FbU_struts_ action=%2Fasset_publisher%2Fview_content&_101_INSTANCE_FXrpx9qY7FbU_ assetEntryId=4289520&_101_INSTANCE_FXrpx9qY7FbU_type=content) (in Portuguese).

[6] Public hearings 27/08/19 – Process number: 25351.911221/2019-74. Brasília: Brazilian Health Regulatory Agency (ANVISA); 2019 (in Portuguese).

[7] Public hearings 08/08/2019 – Process number: 25351.911221/2019-74. Brasília: Brazilian Health Regulatory Agency (ANVISA); 2019 (http://portal.anvisa.gov.br/ audiencias-publicas#/visualizar/400068) (in Portuguese).

[8] Reid OJ, Hammond D, Tariq U, Burkhalter R, Rynard V, Douglas O. Tobacco use in Canada: patterns and trends, 2019 edition. Waterloo (ON): Propel Centre for Population Health Impact, University of Waterloo; 2019 (https://uwaterloo.ca/tobacco-use-canada/ tobacco-use-canadapatterns-and-trends).

[9] Tobacco and Vaping Products Act 2017. Ottawa (ON): Government of Canada; last amended 9 November 2019 (https://laws-lois.justice.gc.ca/eng/acts/T-11.5/).

[10] Notice of intent – potential measures to reduce the impact of vaping products advertising on youth and non-users of tobacco products. In: Government of Canada [website]. Ottawa (ON): Government of Canada; 2019 (https://www.canada.ca/en/ health-canada/programs/consultationmeasures-reduce-impact-vaping-products-advertising-youth-non-users-tobacco-products/notice-document.html).

[11] Consultation summary: notice of intent – potential measures to reduce the impact of vaping products advertising on youth and non-users of tobacco products. Ottawa:

Health Canada; 2019 (https://www.canada.ca/en/health-canada/programs/consultation-measures-reduce-impactvaping-products-advertising-youth-non-users-tobacco-products/notice-document/summary.html).

[12] Vaping Products Promotion Regulations. Canada Gazette Part 1 2019;153(51) (http://www.gazette.gc.ca/rp-pr/p1/2019/2019-12-21/html/reg1-eng.html).

[13] Food and Drugs Act 1985. Ottawa (ON): Government of Canada; last amended 21 June 2019 (https://laws-lois.justice.gc.ca/eng/acts/F-27/index.html).

[14] Canada Consumer Product Safety Act 2010. Ottawa (ON): Government of Canada; last amended 18 October 2018 (https://laws-lois.justice.gc.ca/eng/acts/C-1.68/).

[15] Consumer Chemicals and Containers Regulations, 2001. Ottawa (ON): Government of Canada; last amended 22 June 2016 (https://laws-lois.justice.gc.ca/eng/regulations/sor-2001-269/index.html).

[16] Non-smokers' Health Act 1985. Ottawa (ON): Government of Canada; last amended 17 October 2018 (https://laws-lois.justice.gc.ca/eng/acts/N-23.6/page-1.html).

[17] Walker K. Tobacco taxes not applicable to e-cigarettes. Canadian Tax Focus 2017;7(3):11–12 (https://www.ctf.ca/ctfweb/EN/Newsletters/Canadian_Tax_Focus/2017/3/170315.aspx).

[18] Siekierska A. B.C. hikes tax on vaping products from 7% to 20% [news story]. In: Yahoo Finance [website]. Sunnyvale (CA): Yahoo; 2019 (https://ca.finance.yahoo.com/news/bc-hikes-tax-onvaping-products-205048362.html).

[19] Fiscal plan: a plan for jobs and the economy 2019–23. Edmonton (AB): Alberta Treasury Board and Finance; 2019 (https://open.alberta.ca/dataset/3d732c88-68b0-4328-9e52-5d3273527204/resource/2b82a075-f8c2-4586-a2d8-3ce8528a24e1/download/budget-2019-fiscal-plan-2019-23.pdf).

[20] Budget 2020: fiscal plan. A plan for jobs and the economy 2020–23. Edmonton (AB): Alberta Treasury Board and Finance; 2020 (https://open.alberta.ca/dataset/05bd4008-c8e3-4c84-949e-cc18170bc7f7/resource/79caa22e-e417-44bd-8cac-64d7bb045509/download/budget-2020-fiscal-plan-2020-23.pdf).

[21] Consider the consequences of vaping. In: Government of Canada [website]. Ottawa (ON): Government of Canada; 2019 (https://www.canada.ca/en/services/health/campaigns/vaping.html).

[22] Consider the consequences of vaping (health information video). In: Government of Canada [website]. Ottawa (ON): Government of Canada; 2019 (https://youtu.be/mGaDhpXHWrQ).

[23] Appendix XI, Table 11.2. Adult tobacco survey smokeless tobacco or e-cigarettes. In: WHO report on the global tobacco epidemic, 2019. Geneva: World Health Organization; 2019 (https://www.who.int/tobacco/global_report/en/).

[24] Appendix XI, Table 11.4. Youth tobacco surveys smokeless tobacco or e-cigarettes. In: WHO report on the global tobacco epidemic, 2019. Geneva: World Health Organization; 2019 (https://www.who.int/tobacco/global_report/en/).

[25] Tobacco Business Act 1988. Seoul: National Assembly of the Republic of Korea; last amended 26 July 2017 (https://elaw.klri.re.kr/eng_service/lawView.do?hseq=45814&lang=ENG).

[26] Pharmaceutical Affairs Act 2007. Seoul: National Assembly of the Republic of Korea; last amended 2 December 2016 (https://elaw.klri.re.kr/eng_service/lawView.do?hseq=40196&lang=ENG).

[27] Enforcement Decree of the National Health Promotion Act. Presidential Decree No. 28071, May 29, 2017. Seoul: National Assembly of the Republic of Korea; 2017 (https://elaw.klri.re.kr/kor_service/lawView.do?hseq=43548&lang=ENG).

[28] National Health Promotion Act 2017. Seoul: National Assembly of the Republic of Korea; last amended 30 December 2017 (https://elaw.klri.re.kr/kor_service/lawView.do?lang=ENG&hseq=48657).

[29] Lee S. Health ministry moves to regulate e-cigarette ads [news story]. In: The Korea Times [website]. Seoul: The Korea Times Co.; 2019 (http://www.koreatimes.co.kr/www/nation/2019/09/119_275619.html).

[30] Lee WB. E-cigarette marketing targeted to youth in the Republic of Korea. Tob Control 2017;26(e2):e140–4. doi:10.1136/tobaccocontrol-2016-053448.

[31] Cigarette warning picture and phrase replacement: all e-cigarettes with pictures symbolizing "carcinogenicity". Seoul: Ministry of Health and Welfare; 2018 (http://www.mohw.go.kr/react/al/sal0301vw.jsp?PAR_MENU_ID=04&MENU_ID=0403&page=4&CONT_SEQ=344802) (in Korean).

[32] Enforcement Decree of the Local Tax Act. Presidential Decree No. 28714, 27 March, 2018. Seoul: National Assembly of the Republic of Korea; 2018 (https://elaw.klri.re.kr/eng_service/lawView.do?hseq=47411&lang=ENG).

[33] Reducing tobacco use through taxation: the experience of the Republic of Korea. Washington (DC): World Bank Group; 2018 (http://documents.worldbank.org/curated/en/150681529071812689/pdf/127248-WP-PUBLIC-ADD-SERIES-WBGTobaccoKoreaFinalweb.pdf).

[34] Intensive crackdown on smoking in the non-smoking area. Seoul: Ministry of Health and Welfare; 2019 (http://www.mohw.go.kr/react/al/sal0301vw.jsp?PAR_MENU_ID=04&MENU_ID=0403&page=1&CONT_SEQ=350874) (in Korean).

[35] E-cigarettes. Seventh report of session 2017–19. London: House of Commons Science and Technology Committee; 2018 (https://publications.parliament.uk/pa/cm201719/cmselect/cmsctech/505/505.pdf).

[36] Attitudes of Europeans towards tobacco and electronic cigarettes. Special Eurobarometer 458. Brussels: European Commission; 2017 (https://ec.europa.eu/commfrontoffice/publicopinion/index.cfm/ResultDoc/download/DocumentKy/79003).

[37] Directive 2014/40/EU of the European Parliament and of the Council of 3 April 2014 on the approximation of the laws, regulations and administrative provisions of the member states concerning the manufacture, presentation and sale of tobacco and related products

and repealing Directive 2001/37/EC text with EEA relevance. Brussels: European Union; 2014 (https://eur-lex.europa.eu/legal-content/EN/TXT/?uri=OJ%3AJOL_2014_127_R_0001).

[38] The Tobacco and Related Products Regulations 2016. Statutory Instruments 2016 No. 507. London: The Stationery Office; 2016 (http://www.legislation.gov.uk/uksi/2016/507/contents/made).

[39] Legal framework governing medicinal products for human use in the EU. In: European Commission [website]. Brussels: European Commission; 2019 (https://ec.europa.eu/health/human-use/legal-framework_en).

[40] The Medicines (Codification Amendments Etc.) Regulations 2002. Statutory Instruments 2002 No. 236. London: HMSO; 2002 (http://www.legislation.gov.uk/uksi/2002/236/pdfs/uksi_20020236_en.pdf).

[41] The Medicines for Human Use (Fees Amendments) Regulations 2006. Statutory Instruments 2006 No. 494. London: HMSO; 2006 (https://www.legislation.gov.uk/uksi/2006/494/contents/made).

[42] The Medicines for Human Use (National Rules for Homeopathic Products) Regulations 2006. Statutory Instruments 2006 No. 1952. London: HMSO; 2006 (http://www.legislation.gov.uk/uksi/2006/1952/contents/made).

[43] The General Product Safety Regulations 2005. Statutory Instruments 2005 No. 1803. London: HMSO; 2005 (http://www.legislation.gov.uk/uksi/2005/1803/pdfs/uksi_20051803_en.pdf).

[44] Anastasopoulou S. Overview of the EU electronic cigarette market. Tabexpo Congress 2019, 11-14 November, Amsterdam, the Netherlands [conference presentation]. In: ECigIntelligence [website]. London: Tamarind Media Limited; 2019 (https://ecigintelligence.com/wp-content/uploads/2019/11/TABEXPO-2019_Amsterdam_Stavroula_Anastasopoulou_ECigIntelligence.pdf).

[45] McNeill A, Brose LS, Calder R, Bauld L. Vaping in England: an evidence update February 2019. A report commissioned by Public Health England. London: Public Health England Publications; 2019 (https://assets.publishing.service.gov.uk/government/uploads/system/uploads/attachment_data/file/821179/Vaping_in_England_an_evidence_update_February_2019.pdf).

[46] Nicotine without smoke: tobacco harm reduction. A report by the Tobacco Advisory Group of the Royal College of Physicians. London: Royal College of Physicians; 2016 (https://www.rcplondon.ac.uk/file/3563/download).

[47] E-cigarettes: balancing risks and opportunities. London: British Medical Association; 2019 (https://www.bma.org.uk/collective-voice/policy-and-research/public-and-population-health/tobacco/e-cigarettes).

[48] McNeill A, Brose L, Calder R, Bauld L, Robson D. Evidence review of e-cigarettes and heated tobacco products 2018. A report commissioned by Public Health England. London: Public Health England Publications; 2018 (https://assets.publishing.service.

gov.uk/government/uploads/system/uploads/attachment_data/file/684963/Evidence_review_of_e-cigarettes_and_heated_tobacco_products_2018.pdf).

[49] Secretary of State for Health and Social Care. The Government response to the Science and Technology Committee's seventh report of the Session 2017–19 on e-cigarettes. London: The Stationery Office; 2018 (https://assets.publishing.service.gov.uk/government/uploads/system/uploads/attachment_data/file/762847/government-response-to-science-and-technologycommittee_s-report-on-e-cig.pdf).

[50] Decision: electronic nicotine delivery systems and electronic non-nicotine delivery systems. In: Conference of the Parties to the WHO Framework Convention on Tobacco Control: sixth session, Moscow, Russian Federation, 13–18 October 2014. Geneva: World Health Organization; 2014 (document FCTC/COP6(9); https://apps.who.int/gb/fctc/E/E_cop6.htm).

[51] Report: progress report on regulatory and market developments on electronic nicotine delivery systems (ENDS) and electronic non-nicotine delivery systems (ENNDS). In: Conference of the Parties to the WHO Framework Convention on Tobacco Control: eighth session, Geneva, Switzerland, 1–6 October 2018. Geneva: World Health Organization; 2014 (document FCTC/COP/8/10; https://www.who.int/fctc/cop/sessions/cop8/FCTC_COP_8_10-EN.pdf?ua=1).

[52] Country laws regulating e-cigarettes. In: Global Tobacco Control [website]. Baltimore (MD): Global Tobacco Control; 2020 (https://www.globaltobaccocontrol.org/e-cigarette_policyscan).

附录 A 适用于 EN&NNDS 的国家或联邦法规

本附录显示了EN&NNDS相关法规（按国家和政策领域划分）。

政策领域	巴西	加拿大
产品分类	EN&NNDS在法规[1]中被称为"电子吸烟设备"。意味着它们被分类为烟草制品。"电子吸烟设备"还包括加热型烟草制品。	《烟草和电子烟产品法》[2]将烟草制品定义为全部或部分由烟草（包括烟叶）制造的产品，以及使用此类产品所需的装置（例如烟草加热设备）。但是，它对电子烟产品的定义非常宽泛，即释放气溶胶供人吸入的装置，以及用于这些装置的物质，无论是否含有烟碱。因此，电子烟产品包括零烟碱电子烟液。
向政府提交上市前通告	不需要向政府提交上市前通告。	不需要向政府提交上市前通告。
政府上市前批准	上市前必须获得政府的批准（注册）。巴西国家卫生监督局（ANVISA）可根据提交的证明产品功效、有效性和安全性的毒理学研究和具体的科学测试，批准任何"电子吸烟设备"的销售注册。	如果销售用于治疗目的电子烟产品，则需要获得政府的上市前批准。
进口、销售和分销	除非已在ANVISA注册的产品，否则EN&NNDS的进口、销售和分销均被禁止。目前，没有制造商提交产品注册申请，因此没有已注册的产品。即使对已注册的产品，也禁止向未成年人销售、供应（即使是免费的）和分销任何电子吸烟设备。	根据各省规定，禁止向18岁以下或19岁以下的青少年销售和供应EN&NNDS（法规中的电子烟产品）。新斯科舍省还禁止未成年人拥有电子烟产品。依据2001年《消费化学品和容器法规》（CCCR）[10]第38条，禁止生产、进口[9]或销售烟碱含量 ≥ 66 mg/mL 的电子烟液。

韩国	英国
ENNDS被视为消费品，而根据《烟草商业法》[3]第2条和第3条ENDS被归类为烟草制品。《国民健康促进法》[4]执行法令第27-2条将ENDS定义为"使用电子设备将含烟碱的溶液或切碎的烟草通过呼吸器官吸入人体而产生与吸烟相同效果"的产品。	根据《欧盟烟草制品指令》（TPD）[5]的监管框架，未做出健康声明的EN&NNDS和ENDS产品被归类为消费品，而做出健康声明的EN&NNDS则被视为医药产品。虽然有一种ENDS产品已获得药品许可[6]，但目前还未上市销售。
不需要向政府提交上市前通告。	2016年5月之前上市的所有设备和电子烟烟液的生产商必须在2016年11月之前向药品和保健品管理局（MHRA）提交通告[7]。新设备和电子烟液产品生产商必须在打算上市其产品的前6个月提交通告。 上市前信息必须通告有关成分（包括加热形式）和释放物对健康的潜在影响的毒理学数据，包括致瘾性[8]。"成分"是相关产品中存在的任何物质或元素，包括纸张、滤嘴、油墨、胶囊、黏合剂和任何添加剂。
不需要政府的上市前批准。	只有用于治疗目的的EN&NNDS才需要政府的上市前批准。
禁止向未成年人（未满19岁）出售ENDS。此禁令不适用于ENNDS。	英格兰和威尔士（自2015年10月1日起）、苏格兰（自2017年4月1日起）禁止18岁以下的人购买"烟碱吸入产品"（ENDS和传统卷烟）[11]。此禁令不适用于ENNDS。 欧洲经济区（EEA）或第三国的零售商在向英国销售烟草或ENDS（或两者）之前，必须按照TPD的要求完成注册流程。而英国零售商计划直接向EEA其他成员国消费者销售时，只需要注册。

政策领域	巴西	加拿大
广告、促销和赞助	EN&NNDS 的广告、促销和赞助均被禁止，除非事先在ANVISA进行了注册。如果已注册，可以合理地预期，对吸烟产品的规定将同样适用，广告和促销被禁止，只允许在销售点展示产品。对烟草赞助也有一些限制。	依据2001年《消费化学品和容器法规》（CCCR）第38条[10]，禁止烟碱含量≥66 mg/mL的电子烟液的广告。此外，如果出现以下情况，电子烟产品广告和促销将被禁止： • 产品对18岁以下的人有吸引力，或产品的外观或功能可能使其对这类人群有吸引力； • 使用生活方式广告、客户评价或代言（包括对人、人物或动物的描绘，无论是真实的还是虚构的），无论以何种方式展示和传播，包括通过包装； • 有可能在品牌元素或名称与个人、实体、事件、活动或永久性设施之间产生关联（赞助），或在与个人、实体、事件、活动或永久性设施相关的宣传材料中直接或间接使用与电子烟产品相关的品牌元素或电子烟产品制造商的名称； • 在永久性设施上展示与电子烟产品相关的品牌元素或电子烟制造商名称，作为设施名称的一部分或以其他方式（如果该设施用于体育或文化事件或活动）； • 以虚假、误导或欺骗性的方式展示； • 可能使人相信使用该产品或其释放物可能对健康有益； • 可能会阻碍戒烟或鼓励重新使用烟草制品。
包装和标识	包装 目前还没有关于EN＆NNDS包装的专门立法或法规。	包装 加拿大正在审议《电子烟产品标签和包装条例》[15]。在拟议法规获得批准并生效之前，适用以下要求： • 烟碱含量≥66 mg/mL 的电子烟液符合2001年的CCCR中"剧毒"分类，并根据2001年CCCR第38条禁止制造、进口、广告或销售； • 烟碱含量为10~65 mg/mL的电子烟液符合"有毒"分类，须遵守2001年CCCR对有毒化学品的所有适用要求；零售销售的单独的贮液容器必须具有儿童防护功能，并按照2001年CCCR的要求进行标识，包括在容器的主显示面上贴有毒危险标志。 加拿大卫生部认为烟碱含量在0.1~9 mg/mL的电子烟液在摄入时有潜在毒性，因此必须遵守2001年 CCCR对"有毒"产品的所有要求，包括使用儿童防护容器。

附录A 适用于EN&NNDS的国家或联邦法规

续表

韩国	英国
在电视、广播和广告牌以及其他户外广告支持物上的ENDS广告是非法的。促销折扣等一些营销形式也被禁止。	ENNDS没有被明确监管。在这种情况下,广告和促销受《保护消费者免受不正当交易法规》的约束,保护消费者免受欺骗或骚扰[12]。 如果获得MHRA批准,对做出健康声明的ENDS向公众进行广告和促销,则只有在被视为非处方药时才是合法的[13]。 以下要求仅适用于含烟碱但不属于医药产品的电子烟液、一次性设备和烟弹: • 禁止ENDS的跨境广告以及电视广播广告; • 禁止以下形式的ENDS广告。在报纸、杂志和期刊、商业分类广告、商业电子邮件和短信(除非明确选择)、营销商的在线活动(事实信息除外)、在线促销营销和付费空间中的在线(展示)广告、付费搜索列表、价格比较网站上的优惠列表、病毒式广告、付费社交媒体投放、广告功能和有针对性的品牌内容、游戏内和应用内广告,以及通过电子方式推送到设备上的广告,或通过网络小工具、会员链接和产品植入发布的广告。 含烟碱但不属于医药产品的电子烟烟液、一次性设备和烟弹允许: • 户外广告,包括数字户外广告; • 公共交通工具上的海报(不离开英国); • 电影院、直邮和传单; • 营销人员和消费者之间的私人定制信件; • 专门针对贸易的媒体; • 非广播媒体中的企业广告; • 非跨境活动赞助。 任何广告都必须: • 确保广告对社会负责; • 不以儿童为目标、特征或诉求对象; • 不将电子烟与烟草制品混淆; • 不做医学声明,不随意做健康声明; • 不误导产品成分或使用场所。 英国公共卫生部认为,"声明电子烟至少比吸烟危害小95%仍然是明确传达相对风险巨大差异的一个好方法,从而鼓励更多的吸烟者从吸烟转为吸电子烟"[14]。
包装 到目前为止,尚无适用于EN&NNDS包装的专门的法律规范。	包装 所有含烟碱的容器(一次性设备、烟弹和电子烟液储液罐)必须具有:儿童防护和防篡改包装,防止破裂和泄漏的保护装置,以及确保重新填充不会泄漏的装置。 含烟碱的电子烟液必须装在以下容器中:容量不超过10 mL的专用可填充容器;一次性电子烟中,容量不超过2 mL的一次性烟弹或储液罐。

政策领域	巴西	加拿大
包装和标识	标签要求 目前尚无EN&NNDS标签的具体法律或法规。但是，对于已注册的EN&NNDS，则有理由假定将适用于烟草制品标签法规，包括强制性图形健康警示占包装主要表面积的50%。	标签要求 电子烟产品必须有以下警示："烟碱具有高度的致瘾性"[16]。
成分和释放物产品法规	目前还没有具体的法律规范适用于EN&NNDS成分和释放物的监管，尚未有EN&NNDS产品注册上市。所有增加烟草产品风味和口感使其更具吸引力的添加剂均被禁止[17]。如果EN&NNDS注册了任何此类产品，则该禁令很可能适用于EN&NNDS。 烟草制品禁止使用所有特征性调味剂[17]。如果EN&NNDS注册了任何此类产品，那么这种禁令很可能也适用于EN&NNDS。	强烈建议（但不是法律要求）电子烟具制造商按照关于电子烟和电子烟具电气系统的ANSI/CAN/UL 8139标准对其产品进行认证，并使用符合CAN/CSA-E62133标准或同等标准的锂离子电池。随产品提供的充电器应由加拿大标准委员会认可的认证机构按照适用的加拿大国家标准进行认证。 电子烟烟液中使用的稀释剂应符合公认的药典规定，因此不应使用已知对人体有毒的溶剂，例如乙二醇或二甘醇。 "成分"是指用于制造烟草制品、电子烟产品或其组件的所有物质，包括在该物质的生产中使用的所有物质，就烟草制品而言，还包括烟叶。 禁用的添加剂包括：氨基酸、咖啡因、着色剂、必需脂肪酸、葡萄糖醛酸内酯、益生菌、牛磺酸、维生素和矿物质营养素。 释放物没有明确规定。但是，加拿大卫生部建议，电子烟烟液热分解产生的有害释放物应尽可能少。 建议（但非法律要求），所有添加到电子烟液中的调味剂应为食品级或纯度更高的，并且调味剂中不应使用已知有吸入风险的物质（例如双乙酰和2,3-戊二酮）不应用于调味剂。以下口味被法律禁止：糖果、甜点、大麻、汽水和能量饮料。
税收	到目前为止没有适用于EN&NNDS税收监管的特定法律规范，尚未有注册上市的EN&NNDS。	EN&NNDS仅适用于一般销售税。

续表

韩国	英国
标签要求 ENDS必须带有占包装主要表面50%的图形健康警示。	标签要求 包装必须具有： • 健康警示："该产品含有高度致瘾物质烟碱"；文字必须是白底黑字，覆盖单元包装及任何容器包装前后面30%区域； • 电子烟液中含量为0.1%及以上的成分清单； • 烟碱含量和每次的输送量； • 批号； • 建议将产品放在儿童接触不到的地方。 随附的宣传页（除非包含在包装上）必须有使用和储存说明，包括在适当的情况下重新填充的说明。MHRA建议，这些信息应包括有关产品存储的适当建议，特别是如何确保电池不发生故障、禁忌证、针对特定风险人群的警示和可能的不良影响、致瘾性和毒性以及生产商的详细联系方式，包括欧盟内部的联系方式。
到目前为止没有适用于EN&NNDS成分和释放物监管的具体的法律规范。 对添加剂没有明确规定。 根据《国民健康促进法》第9-3条[4]，可以使用调味剂，但不得进行广告。	不含烟碱的电子烟烟液，无论是在一次性装置中还是在单独的容器中，均受《通用产品安全条例》[18]的监管。符合以下条件的产品被认为是安全的：①在没有联合王国法律的情况下，符合联合王国部分地区的特定健康和安全规则；②联合王国实施官方欧洲标准的自愿性国家标准；③符合其他特定标准或建议，包括有关部门的产品安全良好操作规范，以及消费者对安全的合理期望。对规定的符合性要求所涵盖的风险和风险类别进行安全性评估。 含烟碱电子烟烟液成分： • 必须是高纯度的； • 在加热或未加热的形式下都不会对人体健康造成危害。 电子烟烟液中烟碱的最大浓度为20 mg/mL。 禁用的添加剂包括：使人产生对健康有益处或健康风险降低印象的维生素或其他添加剂，与能量和活力有关的咖啡因、牛磺酸或其他添加剂和兴奋剂化合物，以及对释放物有着色作用的添加剂。 所有其他添加剂的用量不得显著增加或达到可测量程度地增加产品的毒性、致瘾性或致癌性、致突变性或生殖毒性。 释放物没有特别规定。 口味没有特别规定。
ENDS需缴纳多种税费（国民健康促进税、烟草消费税、地方教育和个人消费税），按1799韩元/毫升烟碱溶液（≈1.5美元）的比例进行征税；此外，还需缴纳24韩元/20烟弹（≈0.02美元）的废弃物处理税和10%的增值税（VAT）[19]。	EN&NNDS的增值税税率为20%。

政策领域	巴西	加拿大
室内场所使用	在所有封闭的公共区域包括飞机和公共交通工具，均禁止使用EN&NNDS。封闭的公共区域被定义为可供公众进入或共同使用的公共或私人场所，其任何一面均全部或部分被墙壁、隔板、屋顶、遮阳篷或覆盖物（不论是永久性或临时性的）封闭[20]。	《非吸烟者健康法》[21]禁止在联邦监管的工作场所（如银行、渡轮、客机和联邦政府办公室）室内使用EN&NNDS。大多数省份都禁止在禁止吸烟的场所使用EN&NNDS。
不受商业利益影响	对于保护公众健康不受EN&NNDS行业商业利益影响，没有明确规定。	对于保护公众健康不受EN&NNDS行业商业利益影响，没有明确规定。
监督和监测	到目前为止，没有具体的法律规范适用于EN&NNDS的监督和监测，尚无EN&NNDS注册上市。如果出现违规行为，例如这些产品的商业化、广告以及非法进口，可以通过ANVISA的服务渠道进行投诉。	根据《加拿大消费品安全法》（CCPSA），用于商业用途的电子烟产品的销售商、分销商、进口商、制造商和供应商必须向加拿大卫生部报告和上述产品的供应商报告涉及此类产品的健康或安全事件[24]。根据不同的类型，事故必须在被要求报告人得知事故发生之后的2~10日内报告。 生产、进口、广告、销售或测试用于商业用途的电子烟产品的任何人必须准备和保存文件，说明他们从谁那里获得产品或向谁出售产品，以及其他数据。如果政府提出要求，他们必须提交记录。此要求的目的是，在必须解决危险的情况下，帮助提高不合规产品在供应链中的可追溯性[25]。根据CCPSA，加拿大卫生部有权下令召回产品和采取其他措施，并下令对产品进行测试或研究。 加拿大卫生部对违规产品采取的执法行动取决于与违规行为相关的风险程度。

续表

韩国	英国
在禁止吸烟的场所禁止使用ENDS，但不包括ENNDS。在医疗机构和教育机构（大学除外）完全禁止使用ENDS。在所有其他公共场所和工作场所，只有在指定的禁烟区才禁止吸烟和使用ENDS。	尚无全面的法律来规范在工作场所和公共场所ENDS或ENNDS的使用。每个场所的所有者或管理者可自行决定实施限制。虽然没有官方统计数据，但似乎大多数医院和交通工具都禁止在室内使用电子烟。 根据国际航空运输协会的建议，航空公司严格禁止在飞机上使用电子烟[22]。英格兰公共卫生部发布了关于在公共场所和工作场所使用EN&NNDS的指南，强调吸烟和使用电子烟之间的区别。该指南指出，"对电子烟采取更有利的方法可能是适当的，使其成为比吸烟更容易的选择"[23]。该指南还建议，政策应以对非吸烟者造成危害的证据为基础，风险评估应以证据为依据。该指南表明，二手气溶胶对非吸烟者造成的健康风险极低，根据国际同行评审的证据，禁止在室内使用EN&NNDS的理由不充分。
对于保护公众健康不受EN&NNDS行业商业利益影响，没有明确规定。	对于保护公众健康不受EN&NNDS行业商业利益影响，没有明确规定。
目前还没有专门针对EN&NNDS监督和监测的法律规范，但政府正通过现有的消费者风险监测系统密切监测可能与使用EN&NNDS有关的肺部疾病病例[26]。	消费者和医疗保健专业人员可以通过黄卡计划向MHRA报告副作用和产品安全问题。该计划记录医疗专业人员、制造商或公众报告的药品可疑不良反应。其中包括ENDS和电子烟液。2015年1月1日至2017年10月20日期间共收到37份疑似电子烟不良反应报告，同一报告期内还收到263份与烟碱替代疗法相关的疑似药物不良反应报告。最常报告的不良反应与胃肠不适和呼吸问题有关。

参考文献[1]

[1] Resolution of the Collegiate Board – RDC number 46, 28 August 2009. Prohibits the sale, import and advertising of any electronic smoking devices, known as electronic cigarettes. Brasília: Ministry of Health, Brazilian Health Regulatory Agency (ANVISA); 2009 (http://portal.anvisa.gov.br/documents/10181/2718376/RDC_46_2009_COMP.pdf/2148a322- 03ad-42c3-b5ba-718243bd1919) (in Portuguese).

[2] Tobacco and Vaping Products Act 2017. Ottawa (ON): Government of Canada; last amended 9 November 2019 (https://laws-lois.justice.gc.ca/eng/acts/T-11.5/).

[3] Tobacco Business Act 1988. Seoul: National Assembly of the Republic of Korea; last amended 26 July 2017 (https://elaw.klri.re.kr/eng_service/lawView.do?hseq=45814&lang=ENG).

[4] National Health Promotion Act 2017. Seoul: National Assembly of the Republic of Korea; last amended 30 December 2017 (https://elaw.klri.re.kr/kor_service/lawView.do?lang=ENG&hseq=48657).

[5] Directive 2014/40/EU of the European Parliament and of the Council of 3 April 2014 on the approximation of the laws, regulations and administrative provisions of the member states concerning the manufacture, presentation and sale of tobacco and related products and repealing Directive 2001/37/EC text with EEA relevance. Brussels: European Union; 2014 (https://eur-lex.europa.eu/legal-content/EN/TXT/?uri=OJ%3AJOL_2014_127_R_0001).

[6] e-Voke 10mg & 15Mg electronic inhaler. PL 42601/0003-4. London: Medicines & Healthcare Products Regulatory Agency; 2019 (https://mhraproductsprod.blob.core.windows.net/docs-20200224/56f25daab2a2968139bc37075e194d1a5f12b33f).

[7] Medicines and Healthcare Products Regulatory Agency. In: GOV.UK [website]. London: Government Digital Service (GDS); 2020 (https://www.gov.uk/government/organisations/medicines-and-healthcare-products-regulatory-agency).

[8] The Tobacco and Related Products Regulations 2016. Statutory Instruments 2016 No. 507. London: The Stationery Office; 2016 (https://assets.publishing.service.gov.uk/government/uploads/system/uploads/attachment_data/file/440989/SI_tobacco_products_acc.pdf).

[9] Customs Notice 18-05 ffice; 2016 (https://assets.publishing.service.gov.uk/governmen Products Act (TPVA). In: Canada Border Services Agency [website]. Ottawa (ON): Canada Border Services Agency; 2018 (https://www.cbsa-asfc.gc.ca/publications/cn-ad/cn18-05-eng.html).

[10] Consumer Chemicals and Containers Regulations, 2001. Ottawa (ON): Government of

---

1 所有链接均于 2020 年 3 月 18 日访问。

[11] The Nicotine Inhaling Products (Age of Sale and Proxy Purchasing) Regulations 2015. Statutory Instruments 2015 No. 895. London: The Stationery Office; 2015 (http://www.legislation.gov.uk/uksi/2015/895/pdfs/uksi_20150895_en.pdf).

[12] Marketing and advertising: the law. In: GOV.UK [website]. London: Government Digital Service (GDS); 2019 (https://www.gov.uk/marketing-advertising-law/regulations-thataffect-advertising).

[13] Advertise your medicines: how to comply with the requirements on promoting medicines to the public and to prescribers and suppliers of medicines. In: GOV.UK [website]. London: Government Digital Service (GDS); 2020 (https://www.gov.uk/guidance/advertise-yourmedicines#advertise-to-the-public).

[14] McNeill A, Brose L, Calder R, Bauld L, Robson D. Evidence review of e-cigarettes and heated tobacco products 2018. A report commissioned by Public Health England. London: Public Health England Publications; 2018 (https://assets.publishing.service.gov.uk/government/uploads/system/uploads/attachment_data/file/684963/Evidence_review_of_e-cigarettes_and_heated_tobacco_products_2018.pdf).

[15] Vaping Products Labelling and Packaging Regulations. Canada Gazette Part 1 2019;153(25) (http://gazette.gc.ca/rp-pr/p1/2019/2019-06-22/html/reg4-eng.html).

[16] List of health warnings for vaping products. In: Government of Canada [website]. Ottawa (ON): Government of Canada; 2019 (https://www.canada.ca/en/health-canada/services/smoking-tobacco/vaping/product-safety-regulation/list-health-warnings-vaping-products.html).

[17] The General Product Safety Regulations 2005. Statutory Instruments 2005 No. 1803. London; HMSO: 2005 (http://www.legislation.gov.uk/uksi/2005/1803/pdfs/uksi_20051803_en.pdf).

[18] Resolution of the Collegiate Board uk/uksi/2005/1803/pdfs/uksi_20051803ides for the maximum limits of tar, nicotine and carbon monoxide in cigarettes and the restriction of the use of additives in tobacco products derived from tobacco, and other measures. Brasfndon Ministry of Health, Brazilian Health Regulatory Agency (ANVISA); 2012 (http://bvsms.saude.gov.br/bvs/saudelegis/anvisa/2012/rdc0014_15_03_2012.pdf) (in Portuguese).

[19] Reducing tobacco use through taxation: the experience of the Republic of Korea. Washington (DC): World Bank Group; 2018 (http://documents.worldbank.org/curated/en/150681529071812689/pdf/127248-WP-PUBLIC-ADD-SERIES-WBGTobaccoKoreaFinalweb.pdf).

[20] Decree number 8.262, of 31 May 2014. Amends Decree number 2.018, of 1 October 1996, which regulates Law number 9.294, of 15 July 1996. Brasnumber 2.018, of rs Subsection, Presidency of the Republic Civil House; 2014 (http://www.planalto.gov.br/ccivil_03/_Ato2011-2014/2014/Decreto/D8262.htm) (in Portuguese).

[21] Non-smokershe Republic Civil House; 2014 (http://www.planalto.gov.br/ccivil_03/_ October 2018 (https://laws-lois.justice.gc.ca/eng/acts/N-23.6/page-1.html).

[22] Cabin operations safety best practices guide, 6th edition. Montreal (QC): International Air Transport Association (IATA); 2016.

[23] Use of e-cigarettes in public places and workplaces: advice to inform evidence-based policy making. London: Public Health England Publications; 2016 (https://assets.publishing.service.gov.uk/government/uploads/system/uploads/attachment_data/file/768952/PHE-adviceon-use-of-e-cigarettes-in-public-places-and-workplaces.PDF).

[24] Industry guide on mandatory reporting under the Canada Consumer Product Safety Act – Section 14 "Duties in the event of an incident". Ottawa (ON): Government of Canada; 2018 (https://www.canada.ca/en/health-canada/services/consumer-product-safety/legislation-guidelines/acts-regulations/canada-consumer-product-safety-act/industry/guide-mandatory-reporting-section-14.html).

[25] Guidance on preparing and maintaining documents under the Canada Consumer Product Safety Act (CCPSA) paring and maintainingON): Government of Canada; 2011 (amended 2012) (https://www.canada.ca/en/health-canada/services/consumer-product-safety/legislation-guidelines/guidelines-policies/guidance-preparing-maintaining-documentsunder-canada-consumer-product-safety-act-section-13.html).

[26] What is CISS? In: Korea Consumer Agency: Consumer Injury Surveillance System [website]. Chungbuk: Korea Consumer Agency; 2020 (https://www.ciss.go.kr/english/contents.do?key=595).

附录 B　加拿大各省辖区的其他法规要求

本附件显示了加拿大各省对电子烟碱和非烟碱传输系统（EN&NNDS）的具体规定。

省	适用法律	主要规定
不列颠哥伦比亚省	《烟草和电子烟产品控制法》，2016年9月1日生效[1]。	根据该法律，禁止以下行为： • 在所有封闭的公共场所（包括所有公共和私立学校场地、工作场所和医疗保健机构，指定区域除外）使用EN&NNDS； • 向未成年人（19岁以下）销售和供应； • 在禁止烟草销售的地方销售； • 商店（包括免税店）内的任何促销活动，但在销售点（POS）显示供应情况和价格的促销活动除外； • 除禁止未成年人进入的POS外的所有POS展示； • 仅限成人场所（包括免税商店）的自动售货机。
曼尼托巴省	《非吸烟者健康防护和电子烟产品法案》，2015年11月5日获皇家批准[2]。	根据该法律，禁止以下行为： • 在封闭的公共场所和目前禁止吸烟的其他场所使用EN&NNDS，包括工作场所和载有一名以上乘客的工作车辆（以下场所不在禁止使用EN&NNDS之列：主要销售EN&NNDS的场所以及酒店房间和集体生活设施中的定制吸烟/电子烟室）； • EN&NNDS广告和促销（适用于烟草制品）； • 向未成年人（18岁以下）销售和供应。
新不伦瑞克省	《无烟场所法》，2011年；《烟草和电子烟销售法》，修订案于2018年11月10日生效[3]。	该法律的规定包括以下禁令： • 在禁止吸烟的场所使用EN&NNDS，包括封闭的公共场所、工作场所、餐厅和酒吧，以及有16岁以下人员在场的车辆（部分酒店客房和私人住宅可豁免）； • 向未成年人（19岁以下）销售和供应； • 在禁止烟草销售的地方销售EN&NNDS； • 室内和室外广告和促销材料，即使是在电子烟商店内。
纽芬兰和拉布拉多省	《无烟环境法》[4]和《烟草和电子烟产品管制法》（修订案）[5]，于2018年10月17日生效。	禁止以下行为： • 向未成年人销售； • 在禁止烟草销售的地方销售； • POS促销、产品和宣传材料（不能在店内和店外可见）； • 限制店内招牌； • 如果电子烟商店的唯一业务是销售电子烟产品，则允许其经营。 处理方式：将EN&NNDS参照燃烧型卷烟监管，禁止在公众通常可进入的场所使用EN&NNDS，包括工作场所、私人俱乐部、经许可的餐馆、公交车候车亭以及医疗保健和教育机构。 年轻人：禁止在载有16岁以下未成年人乘客的机动车辆内使用EN&NNDS，并禁止向19岁以下的人销售EN&NNDS和其他电子烟产品。 传播和广告：目前对EN&NNDS广告没有限制。电子烟商店可以向消费者提供有关电子烟的推荐和健康信息。

续表

省	适用法律	主要规定
新斯科舍省	《无烟场所法》（修订案）[6]和《烟草准入法》（修订案）[7]，于2015年5月31日生效。	禁止以下行为： • 在禁止吸烟的场所使用EN&NNDS，包括封闭的公共场所、工作场所、教育和娱乐设施、餐厅和酒吧，以及有19岁以下人员在场的车辆； • 向未成年人（19岁以下）销售和供应； • 未成年人持有； • 在药房销售； • 未按要求展示年龄限制标识； • POS促销，但电子烟商店内除外。
安大略省	2015年《电子烟法案》由2017年《安大略无烟法》(2019年最后一次修订)[8]取代。	禁止以下行为： • 在禁止吸烟的场所使用EN&NNDS，包括封闭的公共场所和工作场所、集体生活设施、公共交通工具或在载有两个或以上工作人员的工作车辆（酒店客房和住宅设施中的指定区域除外）； • 向未成年人（19岁以下）销售和供应； • 向医疗保健机构和药房供应或提议供应； • 在禁止烟草销售的地方进行销售（监管机构可给予某些豁免）； • 在户外、POS（只允许标明价格和供应情况）以及允许儿童进入的场所促销； • 在销售烟草或烟草相关产品的场所或地点有对儿童可见的产品展示。
魁北克省	《烟草控制法》，于2015年11月26日生效[9]。	根据该法律，禁止以下行为： • 在禁止吸烟的场所使用EN&NNDS，包括封闭的公共场所、工作场所、教育和娱乐设施、餐厅和酒吧，以及有16岁以下人员在场的车辆； • 向未成年人出售和供应； • 在禁止烟草销售的地方销售； • 户外标牌，商店只允许显示供应情况和价格。 适用于烟草促销的所有禁令也适用于EN&NNDS。 卷烟中禁用的调味剂不适用于电子烟烟液。
爱德华王子岛	《烟草和电子烟设备销售和准入法》[10]和规定[11]，于2015年10月1日生效。	禁止以下行为： • 在禁止吸烟的场所使用EN&NNDS，包括封闭的公共场所、工作场所、教育和娱乐设施、餐厅和酒吧，以及有19岁以下人员在场的车辆； • 向19岁以下的未成年人销售和供应，以及19岁以下的未成年人购买； • 在禁止烟草销售的地方销售； • POS促销和在零售场所外可见的促销； • 只有在不允许19岁以下人员进入的情况下，电子烟商店才可以展示电子烟； • 任何对这些设备的特性、健康影响和健康危害具有误导性的广告。

参 考 文 献[1]

[1] Tobacco and Vapour Products Control Act, RSBC 1996, c 451. Victoria (BC): Legislative Assembly of British Columbia; 1996 (amended 2018) (http://canlii.ca/t/53gn3).

[2] The Non-Smokers Health Protection and Vapour Products Act. Winnipeg (MB): Legislative Assembly of Manitoba; 1990 (amended 2019) (http://canlii.ca/t/53ncl).

[3] Smoke-free Places Act, RSNB 2011, c 222. Fredericton (NB): Legislative Assembly of New Brunswick; 2011 (amended 2019) (http://canlii.ca/t/53l69).

[4] Smoke-free Environment Act, 2005, SNL 2005, c S-16.2 (as amended). St John's (NL): Newfoundland and Labrador House of Assembly; 2018 (http://canlii.ca/t/53gdz).

[5] Tobacco and Vapour Products Control Act, SNL 1993, c T-4.1. St John53gdz).: Newfoundland and Labrador House of Assembly; 1993 (amended 2018) (http://canlii.ca/t/53gf9).

[6] Smoke-free Places Act (amended). Chapter 54 of the Acts of 2007. Halifax (NS): Nova Scotia Legislature; 2007 (https://nslegislature.ca/legc/bills/60th_2nd/3rd_read/b006.htm).

[7] Tobacco Access Act, SNS 1993, c 14. Chapter 14 of the Acts of 1993. Halifax (NS): Nova Scotia Legislature; 1993 (https://www.canlii.org/en/ns/laws/stat/sns-1993-c-14/latest/sns-1993-c-14.html).

[8] Smoke-Free Ontario Act, 2017, SO 2017, c 26, Sch 3. Toronto (ON): Legislative Assembly of Ontario: 2017 (http://canlii.ca/t/53m3f).

[9] Tobacco Control Act, 2015, c. 28, s. 1. Quebec City (QB): National Assembly of Quebec; 2015 (http://legisquebec.gouv.qc.ca/en/ShowDoc/cs/L-6.2).

[10] Tobacco and Electronic Smoking Device Sales, RSPEI 1988, c T-3.1. Charlottetown: Prince Edward Island Government; amended 2020 (http://canlii.ca/t/548v9).

[11] Tobacco and Electronic Smoking Device Sales and Access Regulations, PEI Reg EC414/05. Charlottetown: Prince Edward Island Government; 2020 (http://canlii.ca/t/548vb).12.

---

1 所有链接均于 2020 年 3 月 18 日访问。

HEATED TOBACCO PRODUCTS
A BRIEF

Abstract

Heated tobacco products (HTPs) are tobacco products that produce an emission containing nicotine and other chemicals, which is then inhaled by users. HTPs are a re-emerging class of tobacco products marketed as so-called potentially reduced-exposure products, or even as modified-risk tobacco products. Currently there is insufficient evidence to conclude that HTPs are less harmful than conventional cigarettes. In fact, there are concerns that while they may expose users to lower levels of some toxicants than conventional cigarettes, they also expose users to higher levels of other toxicants. It is not clear how this toxicological profile translates into short- and long-term health effects. The Conference of the Parties to the WHO Framework Convention on Tobacco Control (WHO FCTC) recognizes HTPs as tobacco products and therefore considers them to be subject to the provisions of the WHO FCTC.

Keywords

HEATED TOBACCO PRODUCTS

EMISSIONS

EFFECTS ON HEALTH

TOBACCO

WHO FCTC

REGULATION

Acknowledgements

This brief was written by Armando Peruga, Consultant, WHO Regional Office for Europe, with contributions from Ranti Fayokun, Scientist, WHO headquarters, Kristina Mauer-Stender, Programme Manager, Angela Ciobanu, Technical Officer, and Elizaveta Lebedeva, Consultant, Tobacco Control Programme, Division of Noncommunicable Diseases and Promoting Health through the Life-course, WHO Regional Office for Europe.

The authors would also like to thank Bente Mikkelsen, Director, Division of Noncommunicable Diseases and Promoting Health through the Life-course, WHO Regional Office for Europe, for her overall leadership and support for the development of this briefing.

The publication was made possible by funding from the Government of Germany.

Introduction

Processed tobacco in heated tobacco products (HTP)[1] is heated without reaching ignition to produce an emission containing nicotine and other chemicals, which is then inhaled by users. HTPs are a re-emerging class of tobacco products marketed as so-called potentially reduced-exposure products, or even as modified-risk tobacco products.

This class of products is defined as re-emerging because at the time of this brief, HTPs were conceptually and technologically an evolution of similar products tobacco companies marketed through the 1980s and 1990s. Back then, the precursors of these products were unsuccessful, and their sale discontinued. The emerging HTPs, however, are expected to capture a significant market share.

Total sales for HTPs in 2016 were US$ 2.1 billion, and they are expected to reach US$ 17.9 billion by 2021 (1). They stand a better chance of profitable marketing today because the tobacco industry is riding partly on the popularity of electronic nicotine and non-nicotine delivery systems (EN&NNDS) in some countries. Although an entirely different class of product, EN&NNDS have changed social norms and perceptions about conventional cigarette-smoking and the use of devices to deliver nicotine in many countries.

Only a few HTPs have been marketed so far (1). Japan Tobacco International (JTI) introduced the Ploom in 2013 jointly with a firm known as Pax Labs, which has continued independently to market the PAX brand. JTI relaunched the Ploom independently in 2016. Philip Morris International (PMI) launched IQOS (I Quit Ordinary Smoking) in 2014. British American Tobacco (BAT) first marketed iFuse in Romania in 2015. Later, BAT marketed Glo in Asia. The Korea Tobacco and Ginseng Corporation (KT&G) is the latest incomer to the HTP market with the lil HTP in 2017 (2). Currently, HTPs are marketed in about 40 countries and IQOS is present in most of them.

There is not much information about the prevalence of HTP use and less about its trends. In Japan, 0.3% of the population aged 15–69 years reported using IQOS in the last 30 days (current use) in 2015 (3). Two years later, this figure was 3.6%. In 2017, 1.2% were currently using Ploom and 0.8% Glo (4). These figures are not mutually exclusive. In Italy, 1.4% of the population aged ≥ 15 years tried IQOS in 2017. Overall, 1.0% of never-smokers, 0.8% of ex-smokers and 3.1% of current cigarette smokers had tried IQOS (5). In Germany, 0.3% of current smokers and recent ex-smokers aged 14 years or more currently used HTPs in 2017 (6). In Great Britain, 1.7% of adults had tried or were using HTPs in 2017, but only 13% of them had been using it daily (7). Three months after the introduction of IQOS in the Republic of Korea in 2017, 3.5% of young adults aged 19–24 years were current users, although all of them also used conventional cigarettes and EN&NNDS (8).

Smoking, the traditional way of extracting nicotine by burning tobacco, results in smoke containing thousands of compounds, many of which are harmful to health. HTPs are based on the principle that burning tobacco is unnecessary to liberate nicotine. In smoking, aerosolizing nicotine is achieved by igniting tobacco, reaching temperatures of up to 900 ºC in the burning cone, but a similar release is attained in HTPs by the volatilization and even pyrolysis (9) of tobacco at temperatures of around 350 ºC, although in some products it may reach up to 550 ºC (10). The lower temperature at which nicotine is volatilized is expected to expose the user to emissions that have fewer toxicants and in smaller amounts than in conventional cigarette smoking. The essential difference between HTPs and EN&NNDS is that the former uses tobacco leaf while the latter does not.

This brief provides a summary of existing evidence of the ingredients, emissions and health effects of HTPs, with a review of policy options for regulation.

---

[1] The tobacco industry calls HTPs "heat-not-burn" products.

Types of HTPs

There are four types of HTPs, depending on how tobacco is heated to deliver nicotine to the user's lungs *(11)*. The first is a cigarette-like device with an embedded heat source that can be used to aerosolize nicotine. The heat is provided by a pressed carbon-tip heat source located at the end of the product, which must be lighted like a conventional cigarette with a standard match or lighter (Fig. 1). Once lit, heat is transferred from the carbon tip to the tobacco, which is not in contact. The resulting temperature of about 350 °C generates an emission infused with nicotine that is inhaled through the mouthpiece. No electrical system is used. After use, the product needs to be extinguished and discarded *(12)*.

FIG. 1. HTP type 1

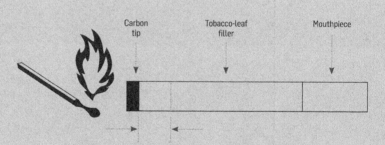

The second type uses an external heat source to aerosolize nicotine from specially designed cigarettes. This is the basic design of IQOS *(13)* (Fig. 2) and Glo *(14)*. The tobacco used in PMI's HTP is apparently not typical tobacco cut-filler but rather a reinforced web of cast-leaf tobacco (a type of reconstituted tobacco) that includes 5–30% by weight of compounds that form emissions, such as polyols, glycol esters and fatty acids. In IQOS, the tobacco is heated by a blade in the heater device inserted into the end of the heat stick (or tobacco-containing element) so that the heat dissipates through the tobacco plug on a puff. The emission then passes through a hollow acetate tube and a polymer film filter on the way to the mouth. BAT describes its Glo product as a heating tube consisting of two separately controlled chambers that are activated by a button on the device to reach the operating temperature (240 °C) within 30–40 seconds.

FIG. 2. HTP type 2

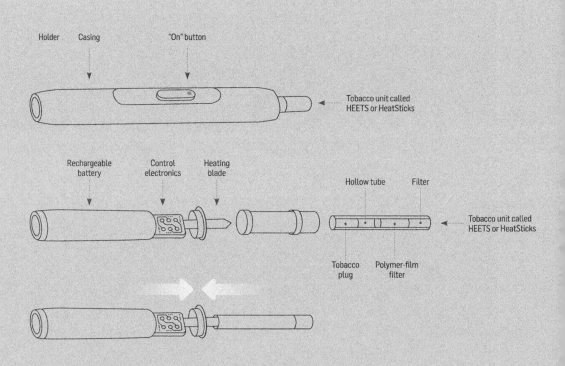

Types of HTPs

A third type uses a heated sealed chamber like a micro-oven (Fig. 3). A battery supplies the power to heat the chamber that transfers the heat through physical contact to any material the user places inside. The user must fill the micro-oven with the grounded tobacco leaf to aerosolize nicotine. The emission is then inhaled by the user through the mouthpiece. This is how dry-herb or loose-leaf vaporizers, such as Pax, work *(15)*. Unlike the other HTPs, the manufacturer does not provide or recommend any of the materials to fill the chamber of the liquid insert.

A fourth type uses a technology similar to EN&NNDS to derive flavour elements from small amounts of tobacco. BAT's iFuse product *(16)* appears to be a hybrid ENDS–tobacco product in which the emission is passed over tobacco to heat it and pick up the flavour and is then inhaled by the user. The JTI Ploom TECH operates in a similar manner *(17)*.

FIG. 3. HTP type 3

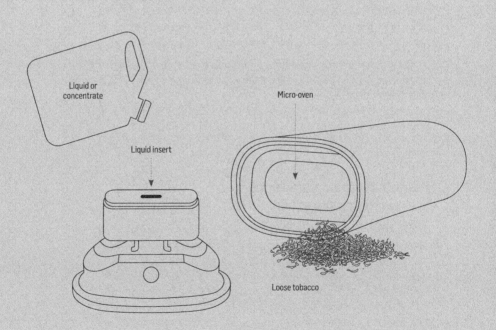

HTP emission content

Nicotine delivery

The mainstream emission from IQOS seems to deliver less nicotine per stick than a conventional cigarette. In studies, nicotine in mainstream emission ranged from 57–83% of that of a reference cigarette. Glo and iFuse deliver less nicotine than IQOS (19–23% of that of a conventional cigarette). HTPs deliver more nicotine than early generations of ENDS [18]. No comparison is available for third-generation ENDS. Nicotine delivery measurements were similar in the tobacco industry and independently funded studies [18].

Studies with humans that measured plasma nicotine levels after use of HTPs indicate that nicotine delivery of HTPs varies by brand but is always lower than that supplied by a conventional cigarette, except for IQOS. Nicotine delivered by HTPs attained peak concentration in plasma as quickly as with conventional cigarettes [18].

Potentially toxic substances in mainstream emission

HTPs' emission contains almost the same number of harmful and potentially harmful compounds (HPHCs) than conventional cigarette smoke, although in some cases at a lower level. A systematic review of published peer-reviewed papers shows that the levels of analysed toxicants were at least 62% lower than in cigarette smoke and particulate matter (PM) was 75% lower than in conventional cigarette smoke [18]. Both tobacco-industry and independently funded studies, including some government institutions in Germany [19], the Netherlands [20] and the United Kingdom [21], found lower levels of toxicants in HTP emission than in cigarette smoke. The independent studies nevertheless reported less tar but more tobacco-specific nitrosamines and, apparently, acetaldehyde, acrolein and formaldehyde than industry-affiliated studies [18].

The finding of a lower level of toxicants in HTP emission must be qualified by the following caveats.

› The number of toxicants measured so far in peer-review articles does not cover the full range of HPHCs of interest. For example, PMI reported in its submission to the United States Food and Drug Administration (FDA) on 40 of the 93 HPHCs in IQOS mainstream emission recommended by the agency. The levels of the missing 53 HPHCs, of which 50 are carcinogenic, are unknown *(22)*.

› The reports submitted by PMI to the FDA include levels of 57 other constituents that are not included in the FDA's list of HPHCs. The level of 56 of them was higher in IQOS emissions than in conventional cigarettes. Their levels were double those in the reference conventional cigarettes for 22 compounds and more than 10 times higher for seven. It appears that IQOS reduces exposure to some toxicants but elevates exposure to other substances. A number of these substances belong to chemical classes that are known to have significant toxicity, but in general, there is limited information on the toxicity of many of them *(22)*.

Potentially toxic substances in side-stream and second-hand emission

Like conventional cigarettes, but unlike EN&NNDS, analysed HTPs generate a side-stream emission. Three studies (one independently funded and two affiliated to the tobacco industry) reported the levels of some HPHCs in IQOS and Glo. All of them found that formaldehyde and acetaldehyde were present in the second-hand emission, although at a level about 10–20 times lower than in cigarette smoke, respectively. Only the independent study found PM and acrolein in the second-hand emission; in this study, PM was about four times lower than in cigarette smoke and acrolein about 50 times lower *(18)*. Consequently, the evidence suggests that second-hand emission from HTPs expose bystanders to quantifiable levels of PM and key toxicants but at a lower level than from second-hand smoke of combustible tobacco products.

Effects of HTP use on health

Nicotine delivery

The nicotine delivery profile of some (but not all) HTPs, particularly IQOS, approximates to that of conventional cigarettes. Some HTPs therefore might be adequate substitutes for cigarettes in the delivery of nicotine, although user satisfaction is reported to be lower than for conventional tobacco products.

Health risks to HTP users from exposure to mainstream emission

There is no available evidence to conclude whether HTP use is associated with any long-term clinical outcome, positive or negative, from exposure to the mainstream emission. One PMI study claimed that IQOS, compared to smoking a conventional cigarette, reduced biomarkers associated with endothelial dysfunction, oxidative stress, inflammation and high-density lipoprotein and cholesterol counts *(23)*. PMI also claimed in the submission to the United States FDA that "human clinical studies have confirmed that clinical markers of ... inflammation show positive changes, similar to those seen following smoking abstinence." A critical review of PMI's data concluded, however, that PMI presented no human clinical data directly from the lung. It also concluded that in human users, there was no evidence of improvement in pulmonary inflammation or pulmonary function in cigarette smokers who switched to IQOS. PMI's claim that smokers who switch to IQOS reduce inflammation and the risk of chronic obstructive pulmonary disease therefore is not supported even by their own data. There are very few independent studies reporting on the short-term effects of HTP use. They indicate some short-term physiopathological effect *(24–26)*.

Health risks from exposure to HTP second-hand emission

There is no available evidence to indicate whether HTP use is associated with any long-term clinical outcomes from exposure to the second-hand emission. HTPs nevertheless generate side-stream emission with ultrafine particles and a number of harmful toxicants, although at a lower level than in conventional cigarettes. A recent study found that a proportion of people exposed to second-hand IQOS emissions experienced short-term symptoms such as sore throat, eye pain and feeling ill *(4)*.

Given that a number of public health organizations, including WHO *(27,28)*, have deemed that no level of side-stream exposure is safe or acceptable, these findings are clearly concerning and merit further study.

Key messages

> HTPs contain tobacco and emit nicotine and other toxicants.

> HTPs generate a mainstream emission and a side-stream emission. Inhaling the mainstream emission exposes HTP users to the toxicants contained in the emission. Bystanders may inhale the side-stream or second-hand emissions.

> Currently there is insufficient evidence to conclude that HTPs are less harmful than conventional cigarettes. In fact, there are concerns that while they may expose users to lower levels of some toxicants than conventional cigarettes, they also expose users to higher levels of other toxicants. It is not clear how this toxicological profile translates into short- and long-term health effects.

Conclusions

Governments should introduce a system for the pre-market assessment of novel tobacco products, including HTPs. Marketing of HTPs should not be permitted unless there is conclusive evidence that compared to conventional cigarettes, the product reduces exposure to harmful and potentially harmful components and reduces health risks.

Governments that cannot prevent the introduction of HTPs in their markets or decide to allow the marketing of HTPs in the absence of such evidence should ensure the tobacco industry cannot claim government authorization of the product as its endorsement.

In addition, HTPs should be taxed similarly to other tobacco products, following the recommendations of the Conference of the Parties to the WHO Framework Convention on Tobacco Control (WHO FCTC) in its decision FCTC/COP8(22) *(29)*. The decision recognized HTPs as tobacco products and therefore considers them to be subject to the provisions of the WHO FCTC. The decision also reminded Parties to prioritize the following measures in accordance with the WHO FCTC and national law:

- prevent the initiation of HTP use;
- protect people from exposure to HTP emissions and explicitly extend the scope of smoke-free legislation to these products in accordance with Article 8 of the WHO FCTC;
- prevent health claims being made about HTPs;
- apply measures regarding advertising, promotion and sponsorship of HTPs in accordance with Article 13 of the WHO FCTC;
- regulate the contents and disclosure of contents of HTPs in accordance with articles 9 and 10 of the WHO FCTC;
- protect tobacco-control policies and activities from all commercial and other vested interests related to HTPs, including interests of the tobacco industry, in accordance with Article 5.3 of the WHO FCTC; and
- regulate, including restrict, or prohibit, as appropriate, the manufacture, importation, distribution, presentation, sale and use of HTPs as appropriate to national laws, taking into account a high level of protection for human health.

Finally, it is important to monitor comprehensively not only market developments, but also the use of HTPs through inclusion of relevant questions in all appropriate surveys.

References[2]

1. Heated tobacco products (HTPs) market monitoring information sheet. In: World Health Organization [website]. Geneva: World Health Organization; 2019 (https://www.who.int/tobacco/publications/prod_regulation/htps-marketing-monitoring/en/).

2. Lee J, Lee S. Korean-made heated tobacco product, "lil". Tob Control 2018;tobaccocontrol-2018-054430. doi:10.1136/tobaccocontrol-2018-054430.

3. Tabuchi T, Kiyohara K, Hoshino T, Bekki K, Inaba Y, Kunugita N. Awareness and use of electronic cigarettes and heat-not-burn tobacco products in Japan. Addiction 2016;111(4):706–13. doi:10.1111/add.13231.

4. Tabuchi T, Gallus S, Shinozaki T, Nakaya T, Kunugita N, Colwell B. Heat-not-burn tobacco product use in Japan: its prevalence, predictors and perceived symptoms from exposure to secondhand heat-not-burn tobacco aerosol. Tob Control 2017;27(e1):e25–33. doi:10.1136/tobaccocontrol-2017-053947.

5. Liu X, Lugo A, Spizzichino L, Tabuchi T, Pacifici R, Gallus S. Heat-not-burn tobacco products: concerns from the Italian experience. Tob Control 2018;tobaccocontrol-2017-054054. doi:10.1136/tobaccocontrol-2017-054054.

6. Kotz D, Kastaun S. E-Zigaretten und Tabakerhitzer: repräsentative Daten zu Konsumverhalten und assoziierten Faktoren in der deutschen Bevölkerung (die DEBRA-Studie) [E-cigarettes and tobacco heaters: representative data on consumer behaviour and associated factors in the German population (the DEBRA study)]. Bundesgesundheitsblatt Gesundheitsforschung Gesundheitsschutz 2018;61(11):1407–14. doi:10.1007/s00103-018-2827-7 [in German].

7. Brose L, Simonavicius E, Cheeseman H. Awareness and use of "heat-not-burn" tobacco products in Great Britain. Tob Regul Sci. 2018;4(2):44–50. doi:10.18001/trs.4.2.4.

8. Kim J, Yu H, Lee S, Paek Y. Awareness, experience and prevalence of heated tobacco product, IQOS, among young Korean adults. Tob Control 2018;27(Suppl. 1):s74–7. doi:10.1136/tobaccocontrol-2018-054390.

9. Davis B, Williams M, Talbot P. iQOS: evidence of pyrolysis and release of a toxicant from plastic. Tob Control 2019;28:34–41. doi:10.1136/tobaccocontrol-2017-054104.

10. Jiang Z, Ding X, Fang T, Huang H, Zhou W, Sun Q. Study on heat transfer process of a heat not burn tobacco product flow field. J Phys Conf Ser. 2018;1064:012011. doi:10.1088/1742-6596/1064/1/012011.

11. O'Connor R. Heated tobacco products. In: WHO study group on tobacco product regulation. Report on the scientific basis of tobacco product regulation: seventh report of a WHO study group. Geneva: World Health Organization; 2019:3–29 (https://apps.who.int/iris/bitstream/handle/10665/329445/9789241210249-eng.pdf?ua=1).

12. Platform 2. In: PMI Science [website]. Neuchâtel: PMI Science; undated (https://www.pmiscience.com/our-products/platform2).

13. Our tobacco heating system: IQOS. In: Philip Morris International [website]. Neuchâtel: Philip Morris International; undated (https://www.pmi.com/smoke-free-products/iqos-our-tobacco-heating-system).

14. Tobacco heating products. In: British American Tobacco. London: British American Tobacco; 2019 (https://www.bat.com/group/sites/UK__9D9KCY.nsf/vwPagesWebLive/DOAWUGNJ#).

15. PAX 3 FAQ. In: Pax Labs [website]. San Francisco (CA): Pax Labs; undated (https://www.pax.com/pages/pax-3-faq).

16. Spencer B. The iFuse "hybrid" cigarette combines e-cig technology with tobacco to improve the flavour of the vapour. Daily Mail online. 23 November 2013 (https://www.dailymail.co.uk/sciencetech/article-3330238/The-iFuse-hybrid-cigarette-combines-e-cig-technology-tobacco-improve-flavour-vapour.html).

---

[2] All weblinks accessed 1 December 2019.

17. Reduced-risk products (RRP). In: Japan Tobacco International [website]. Geneva: Japan Tobacco International; undated (https://www.jt.com/sustainability/our_business/tobacco/rrp/index.html).

18. Simonavicius E, McNeill A, Shahab L, Brose L. Heat-not-burn tobacco products: a systematic literature review. Tob Control 2018;tobaccocontrol-2018-054419. doi:10.1136/tobaccocontrol-2018-054419.

19. Mallock N, Böss L, Burk R, Danziger M, Welsch T, Hahn H et al. Levels of selected analytes in the emissions of "heat not burn" tobacco products that are relevant to assess human health risks. Arch Toxicol. 2018;92(6):2145–9. doi:10.1007/s00204-018-2215-y.

20. Addictive nicotine and harmful substances also present in heated tobacco. In: National Institute for Public Health and the Environment (RIVM) [website]. Bilthoven: National Institute for Public Health and the Environment (RIVM); 2018 (https://www.rivm.nl/en/news/addictive-nicotine-and-harmful-substances-also-present-in-heated-tobacco).

21. Statement on heat not burn tobacco products. London: Food Standards Agency; 2017 (https://cot.food.gov.uk/committee/committee-on-toxicity/cotstatements/cotstatementsyrs/cot-statements-2017/statement-on-heat-not-burn-tobacco-products).

22. St Helen G, Jacob Iii P, Nardone N, Benowitz N. IQOS: examination of Philip Morris International's claim of reduced exposure. Tob Control 2018;27(Suppl. 1):s30–6. doi:10.1136/tobaccocontrol-2018-054321.

23. Lüdicke F, Picavet P, Baker G, Haziza C, Poux V, Lama N et al. Effects of switching to the menthol tobacco heating system 2.2, smoking abstinence, or continued cigarette smoking on clinically relevant risk markers: a randomized, controlled, open-label, multicenter study in sequential confinement and ambulatory settings (part 2). Nicotine Tob Res. 2017;20(2):173–82. doi:10.1093/ntr/ntx028.

24. Leigh N, Tran P, O'Connor R, Goniewicz M. Cytotoxic effects of heated tobacco products (HTP) on human bronchial epithelial cells. Tob Control 2018;27(Suppl. 1):s26–9. doi:10.1136/tobaccocontrol-2018-054317.

25. Biondi-Zoccai G, Sciarretta S, Bullen C, Nocella C, Violi F, Loffredo L et al. Acute effects of heat-not-burn, electronic vaping, and traditional tobacco combustion cigarettes: the Sapienza University of Rome vascular assessment of proatherosclerotic effects of smoking (SUR-VAPES) 2 randomized trial. J Am Heart Assoc. 2019;8(6): e010455. doi:10.1161/jaha.118.010455.

26. Sohal S, Eapen M, Naidu V, Sharma P. IQOS exposure impairs human airway cell homeostasis: direct comparison with traditional cigarette and e-cigarette. ERJ Open Res. 2019;5(1):00159-2018. doi:10.1183/23120541.00159-2018.

27. Policy recommendations on protection from exposure to second-hand tobacco smoke. Geneva: World Health Organization; 2007 (https://www.who.int/tobacco/publications/second_hand/protection_second_hand_smoke/en/).

28. Guidelines for implementation of Article 8. Guidelines on the protection from exposure to tobacco smoke. Geneva: World Health Organization; 2007 (https://www.who.int/fctc/treaty_instruments/adopted/article_8/en/).

29. Conference of the Parties to the WHO Framework Convention on Tobacco Control. Eighth session. Geneva, Switzerland, 1–6 October 2018. Decision. FCTC/COP8(22). Novel and emerging tobacco products. Geneva: World Health Organization; 2018 (https://www.who.int/fctc/cop/sessions/cop8/FCTC__COP8(22).pdf).

ELECTRONIC NICOTINE AND NON-NICOTINE DELIVERY SYSTEMS
A BRIEF

Abstract

Electronic nicotine and non-nicotine delivery systems (EN&NNDS) are a heterogeneous class of products that use an electrically powered coil to heat and turn a liquid into an aerosol, which is inhaled by the user. EN&NNDS are not harmless. Although the consequences for long-term effects on morbidity and mortality have not yet been studied sufficiently, EN&NNDS are not safe for young people, pregnant women and adults who have never smoked. While it is expected that use of EN&NNDS in these groups might increase their health risks, non-pregnant adult smokers who completely and promptly switch from combustible tobacco cigarettes to use of unadulterated and appropriately regulated EN&NNDS alone might reduce their health risks. Member States that decide to regulate EN&NNDS may consider, inter alia: regulating EN&NNDS that make health claims as medicinal products and therapeutic devices; banning or restricting advertising, promotion and sponsorship of EN&NNDS; minimizing health risks to non-users by outlawing the use of EN&NNDS in all indoor spaces or where smoking is prohibited; and limiting the level and number of specific flavours allowed in EN&NNDS to reduce initiation by young people.

Keywords

ELECTRONIC NICOTINE DELIVERY SYSTEMS (ENDS)

ELECTRONIC NON-NICOTINE DELIVERY SYSTEMS (ENNDS)

EN&NNDS CONTENTS

EFFECTS ON HEALTH

ROLE IN SMOKING CESSATION

WHO FCTC

REGULATION

Acknowledgements

This brief was written by Armando Peruga, Consultant, WHO Regional Office for Europe, with contributions from Ranti Fayokun, Scientist, WHO headquarters, Kristina Mauer-Stender, Programme Manager, Angela Ciobanu, Technical Officer, and Elizaveta Lebedeva, Consultant, Tobacco Control Programme, Division of Noncommunicable Diseases and Promoting Health through the Life-course, WHO Regional Office for Europe.

The authors would like to thank Bente Mikkelsen, Director, Division of Noncommunicable Diseases and Promoting Health through the Life-course, WHO Regional Office for Europe, for her overall leadership and support for the development of this briefing.

The publication was made possible by funding from the Government of Germany.

Electronic nicotine and non-nicotine delivery systems

Electronic nicotine and non-nicotine delivery systems (EN&NNDS)[1] are a heterogeneous class of products that use an electrically powered coil to heat and turn a liquid into an aerosol, which is inhaled by the user.

The generation and composition of aerosol during EN&NNDS use or so-called vaping and the subsequent exposure to aerosol substances is determined by four factors:

1_ the e-liquid composition
2_ the materials used in manufacturing the device
3_ the electrical power or wattage used during operation to heat the e-liquid
4_ the puffing topography or inhaling characteristics of the user while using EN&NNDS.

E-liquids always contain carrier liquids (humectants) which comprise 80–90% of the volume of the liquid, some water (10–20% of the volume) and, generally, nicotine and flavours. Propylene glycol and glycerol, the principal carriers used in e-liquids, undergo partial decomposition in contact with the atomizer heating coil, forming several toxicants, including carbonyls. E-liquids may also contain nicotine, a highly addictive substance, that may adversely affect fetal and adolescent brain development.

Heating elements or coils in EN&NNDS are usually made of resistance wires of various metals, such as nickel, or metal alloys, including nichrome (chrome and nickel). Metal parts of the device are sometimes soldered with lead.

To heat and aerosolize the e-liquid, an electrical current from a battery is run through the coil when the EN&NNDS is activated. The temperature reached depends on the electrical power generated, which in turn depends on the amount of energy provided by the battery and the resistance of the coil. The lower the resistance, the more electricity flows through it, and the more heat is generated at the coil. Under normal operation conditions, the e-liquid reaches a temperature of between 100–350 °C.

The inhaling behaviour or puffing topography of users has the following variables: puff volume, depth of inhalation, rate of puffing, and intensity of puffing. These variables determine the amount of aerosol inhaled and how deep it is carried into the respiratory system.

---

[1] This brief follows the terminology of the Conference of the Parties of the WHO Framework Convention on Tobacco Control by referring to electronic nicotine delivery systems (ENDS) for products in which the e-liquid contains nicotine and electronic non-nicotine delivery systems (ENNDS) for those in which it does not. These systems are referred to collectively as electronic nicotine and non-nicotine delivery systems (EN&NNDS). They are popularly known as electronic cigarettes or e-cigarettes. Other sources refer to ENDS as alternative nicotine delivery systems (ANDS).

EN&NNDS use among the population

Proportion of the adult population using EN&NNDS regularly

The proportion of adults who currently (defined as at least once in the last month) used EN&NNDS in the two main world markets for these products, the United States of America and the European Union (EU), was 3.2% in 2018 *(1)* and 2% in 2017 *(2)* respectively. Highest prevalence of use among EU countries in 2018 was in the United Kingdom (England), at 6.2% *(3)*. In New Zealand, 3.8% of adults currently used EN&NNDS in 2017–2018 *(4)*. Additional data from nine countries indicate that in most, no more than 4% of adults used EN&NNDS regularly between 2017 and 2018 *(5)*.

Very few countries have trend data. The proportion of current adult users of EN&NNDS in the United States has remained stable since 2014 (at 3.7%) *(1)* and in the EU since 2015 (2%) *(6)*. In Canada, past 30-day use and daily use of EN&NNDS among adults remained stable in the period between 2013 and 2017 *(7)*. Only New Zealand shows a clear increase of use of EN&NNDS among adults, from 1.4% in 2015–2016 to 3.8% in 2017–2018 and 4.7% in 2018–2019 *(4)*. Most EN&NNDS users are or were smokers.

Proportion of young people using EN&NNDS regularly

Data on the current use of EN&NNDS among young people aged 13–15 years from 22 countries indicate that the proportion using EN&NNDS regularly is higher than among their adult counterparts. Figures for young people ranged from 0.7% in Japan to 18.4% in Ukraine between 2017 and 2019, with a median country value of 8.1% *(8)*.

Between 2008 and 2015, ever using EN&NNDS among young people increased in Poland, New Zealand, the Republic of Korea and the United States, decreased in Canada and Italy, and remained stable in the United Kingdom *(9)*. Current use of EN&NNDS among 11–18-year-olds in the United States increased from 2017 to 2018 *(10)* while remaining stable in the United Kingdom. In 2019, 1.6% of 11–18-year-olds in the United Kingdom used EN&NNDS more than once a week, compared to 1.7% in 2018 *(11)*. In Canada, past 30-day use of EN&NNDS among young people in grades 7–9 was 5.4% in 2016–2017, which was not significantly different from use in 2014–2015 *(12)*. A recent study comparing the change in EN&NNDS use among 16–19-year-olds in Canada, the United Kingdom (England) and the United States between 2017 and 2018 confirmed the increase of EN&NNDS use in Canada and the United States and the stability in the United Kingdom (England) for use during the past 30 days and past week (Table 1) *(13)*.

EN&NNDS use during:	Canada		United States		United Kingdom (England)	
	2017 (%)	2018 (%)	2017 (%)	2018 (%)	2017 (%)	2018 (%)
past 30 days	8.4	14.6	11.1	16.2	8.7	8.9
past week	5.2	9.3	6.4	10.6	4.6	4.6

TABLE 1. Prevalence change of current EN&NNDS use among 16–19-year-olds between 2017 and 2018 in three countries

Source: Hammond et al. *(13)*.

EN&NNDS current use among non-smoking young people

Data from the United States show that in 2017, 0.8% of all 11–18-year-olds who had never smoked a cigarette before were using EN&NNDS regularly (at least once in the last 10 days). In 2018, the proportion increased to 2.4% *(14)*. In the United Kingdom (England), however, 0.8% of young people aged 11–18 years who had never smoked were currently using EN&NNDS *(3)*. Weekly use of EN&NNDS among never-smokers aged 17 and 18 years was 0% in 2016 and 2017 *(15)* and 0.2% in 2018 *(3)*. The recent study comparing the change in EN&NNDS use among 16–19-year-olds in Canada, the United Kingdom (England) and the United States between 2017 and 2018 confirmed the increasing number of never-smokers using EN&NNDS during the past 30 days and the past week in Canada and the United States, while the figure did not change in the United Kingdom (England) (Table 2) *(13)*.

EN&NNDS use during:	Canada		United States		United Kingdom (England)	
	2017 (%)	2018 (%)	2017 (%)	2018 (%)	2017 (%)	2018 (%)
past 30 days	2.3	5.0	2.4	5.9	1.6	1.9
past week	0.8	2.7	1.1	3.0	0.5	0.4

TABLE 2. Prevalence change of current EN&NNDS use among 16–19-year-old never-smokers between 2017 and 2018 in three countries

Source: Hammond et al. *(13)*.

EN&NNDS contents and health effects

EN&NNDS aerosol contents

The aerosol users breathe from EN&NNDS contains numerous potentially toxic substances, in addition to nicotine when included in the e-liquid. The number, quantity and characteristics of potentially toxic substances in the aerosol emitted by EN&NNDS are highly variable and depend on product characteristics (including device and e-liquid features) and how the device is operated by the user. Under typical conditions of use, however, the number and concentrations of potentially toxic substances emitted from unadulterated EN&NNDS are lower than in tobacco smoke, except for some metals.

The main substances in the aerosol that raise health concern are **metals**, such as chromium, nickel, and lead, and **carbonyls**, such as formaldehyde, acetaldehyde, acrolein and glyoxal.

The types and concentrations of **metals** depend on the product features and inhaling patterns of use. Exposure to certain levels of some metals may cause serious health effects, such as diseases of the nervous, cardiovascular and respiratory systems. The number of metals in the aerosol could be greater than in combustible tobacco cigarettes, and in some cases is found at higher concentrations than in cigarette smoke. It is suspected that metals come mostly from the metallic coil used to heat the e-liquid and soldered joints of the device. Metal emissions can largely be prevented through appropriate engineering of devices.

Carbonyl compounds are potentially hazardous to users. Formaldehyde is a human carcinogen, acetaldehyde is possibly carcinogenic to humans, acrolein is a strong irritant of the respiratory system and glyoxal shows mutagenicity. Most carbonyls come from the thermal decomposition of humectants, propylene glycol and glycerol. The number and levels of carbonyls detected in the aerosol are lower than in smoke from combustible tobacco, but even these levels raise health concerns.

Other substances in the aerosol of possible health concern are **particulate matter** and some **flavourings**.

The **particle** count and size in EN&NNDS aerosols do not differ greatly from those found in mainstream combustible tobacco smoke. The composition of the particles nevertheless is dissimilar and likely to have a different health impact. Aerosol particulates from EN&NNDS consist mostly of a mix of aqueous and humectant droplets, whereas particles in combustible tobacco smoke are mostly complex organic constituents that contain known or suspected carcinogens.

Although of health concern, particles from EN&NNDS are therefore expected to have smaller health risks than particles in tobacco smoke.

Certain **flavourings**, such as diacetyl, cinnamaldehyde and benzaldehyde, have been cited as a source of health concerns when heated and inhaled.

When the e-liquid contains **nicotine**, the aerosol contains nicotine. The amount of nicotine inhaled by ENDS users is highly variable and depends on product characteristics (including device and e-liquid characteristics) and how the device is operated. There is substantial evidence that nicotine intake from ENDS among experienced adult ENDS users can be comparable to that from combustible tobacco cigarettes.

Health effects of using EN&NNDS

Scientists are still learning about the long-term health effects of EN&NNDS. Currently, there is insufficient research to determine with certainty whether unadulterated and appropriately regulated EN&NNDS use is associated with cardiovascular, lung or cancer diseases.

The following section describes what currently is known by strength of evidence,[2] as assessed by the National Academies of Sciences, Engineering and Medicine (NASEM) in 2018 *(16)*.

There is **conclusive evidence** that:

- completely substituting EN&NNDS for combustible tobacco cigarettes reduces users' exposure to numerous toxicants and carcinogens present in combustible tobacco cigarettes;
- EN&NNDS devices can explode and cause burns and projectile injuries when batteries are of poor quality, stored improperly or modified by users; and
- intentional or accidental exposure to e-liquids (from drinking, eye contact or dermal contact) can result in adverse health effects, sometimes fatal.

There is **substantial evidence** that:

- ENDS use results in symptoms of nicotine dependence – the risk and severity of nicotine dependence are influenced by the ENDS product characteristics (nicotine concentration, flavouring, device type and brand), but the risk and severity of dependence seem lower for ENDS than from combustible tobacco cigarettes;

---

2   Only conclusions for which the NASEM deems there to be conclusive, substantial and moderate evidence are presented, not evidence with limited, insufficient or no available proof. Conclusive, substantial and moderate scientific evidence allows firm conclusions, firm conclusions with minor limitations, and a general conclusion with limitations, respectively. Strength of evidence refers to the certainty of an association but not necessarily to its magnitude. The strength of using reliable literature reviews, such as the one from NASEM, is that their conclusions are based on the systematic and methodical overview of consolidated evidence at the moment of the review, 2018 in this case. Research on EN&NNDS is developing rapidly, however, which means that some new studies may contradict the conclusions of systematic reviews. Findings of new studies not yet included in reliable systematic reviews are not discussed here, unless they present overwhelming and undisputed evidence.

-) EN&NNDS aerosol can cause some human cells to malfunction – it is not clear what this means in terms of the long-term consequences of chronic use of EN&NNDS, but it is possible that it could increase the risk of some diseases, such as cardiovascular disease, cancer and adverse reproductive outcomes, although the risk is probably lower than from combustible tobacco cigarette smoke; and
-) completely switching from regular use of combustible tobacco cigarettes to EN&NNDS results in reduced short-term adverse health outcomes in several organ systems.

There is **moderate evidence** that:

-) EN&NNDS use increases cough and wheeze in adolescents and is associated with an increase in asthma exacerbations; and
-) the positive and negative health impacts of EN&NNDS use is applicable to the employment of these products in the absence of simultaneous consumption of tobacco products, but a significant proportion of EN&NNDS users, referred to as dual- or poly-users, also smoke tobacco products.

The question is – do EN&NNDS users who continue to smoke have any reduction in health risk? The NASEM review concluded that there is no available evidence on whether long-term e-cigarette use among smokers (dual use) changes morbidity or mortality compared with those who only smoke combustible tobacco cigarettes. Recent evidence, however, suggests that dual users have a greater level of oxidative stress than smokers *(17)* and that adding use of EN&NNDS to smoking may contribute to cardiopulmonary health risks, particularly involving the respiratory system *(18)*.

A note on the cases of lung problems linked to EN&NNDS use in the United States

During the drafting of this brief, the Centers for Disease Control and Prevention (CDC) of the United States reported an outbreak of lung disorders associated with the use of e-cigarettes and vaping *(19)*. As of 7 January 2020, more than 2500 cases had been reported to CDC from 50 states. Almost 60 deaths had been confirmed in 27 states.

CDC has identified vitamin E acetate as a chemical of concern among people with e-cigarette, or vaping, product-use associated lung injury (EVALI). CDC laboratory testing of fluid samples collected from the lungs of 29 patients with EVALI submitted from 10 states found vitamin E acetate in all of the samples. Vitamin E acetate is used as an additive, most notably as a thickening agent in tetrahydrocannabinol-containing e-cigarette, or vaping, products.

Another study concludes that dual users are not reducing exposure to harmful toxicants compared to exclusive cigarette smokers due to their continued smoking *(20)*. A possible explanation is that dual users include a great variety of tobacco- and EN&NNDS-use behaviours, each with different motivations *(21)*. Dual use may not represent only a transitional phase to reduce or quit smoking; this category may also include EN&NNDS users who still rely on smoking to manage their dissatisfaction with the EN&NNDS experience, to circumvent smoke-free policies or simply to comply with social group norms and manage the stigma associated with smoking *(22)*.

Second-hand exposure to EN&NNDS aerosol

EN&NNDS users inhale the aerosol directly from the device and partly exhale it back into the air, which bystanders may then breathe in. As a result, **EN&NNDS use increases airborne concentrations of particulate matter and nicotine in indoor environments compared with background levels** *(16)*. Some studies indicate that some volatile organic compounds are also exhaled into the environment during EN&NNDS use. The concentration of these substances in the air increases with the number of users in confined spaces. **Second-hand exposure to nicotine and particulates is lower from EN&NNDS aerosol compared with combustible tobacco cigarettes** *(16)* but are higher than the smoke-free level recommended by the WHO Framework Convention on Tobacco Control (WHO FCTC) *(23)*.

Health effects of exposure to exhaled aerosol

No available studies have evaluated the health effects of second-hand EN&NNDS exposure, so the risks to health of exposure to exhaled aerosol remain unknown. It is expected, however, to present some health risks for bystanders, although at lower levels than from exposure to second-hand tobacco smoke.

EN&NNDS' role in smoking cessation and initiation

EN&NNDS' role in smoking cessation among adults

The NASEM review concluded that there is insufficient evidence from randomized controlled trials about the effectiveness of ENDS as cessation aids compared with no treatment or approved smoking-cessation treatments *(16)*, although it did not include a recent trial whose results depart from this conclusion *(24)*. Moderate evidence, however, shows that some smokers may successfully quit tobacco by using some types of ENDS frequently or intensively *(16)*, while others experience no difference or are even prevented from quitting *(25)*.

EN&NNDS' role in smoking initiation among young people

There is moderate evidence that young never-smokers who experiment with EN&NNDS are at least twice more likely to experiment with smoking later *(16)*. The data available so far do not, however, prove that this evident association is causal. While some authors believe that ENDS use and smoking are initiated independently of each other as the result of a common latent propensity to risky behaviour, others think that the similarity between ENDS use and smoking facilitates the trajectory from one product to the other within a social learning framework.

The role of flavours in EN&NNDS initiation and use

E-liquids for EN&NNDS are marketed in more than 15 000 unique flavours *(26,27)*. Flavours are classified in two big groups: tobacco flavours, and those that impart a strong non-tobacco smell or taste. The latter are considered so-called characterizing flavours, the main categories of which are menthol/mint, nuts, spices, coffee/tea, alcohol, other beverages, fruit, candy and other sweets *(28)*.

Flavours are one of the most appealing features of EN&NNDS and have been described as the major motivation for ENDS use by young people. They can alter expectations and reward from EN&NNDS, including nicotine effects *(29,30)*. Advertisements on e-liquid containers and vendor websites frequently contain images and descriptions of flavours that convey appealing product sensations *(31)*.

Flavours seem to play a part in promoting the switch from combustible tobacco products to EN&NNDS *(32–34)*. They also play an important role in increasing uptake of EN&NNDS among young people *(35–37)*, noticeably more significantly than among adults *(38)*. The use of flavoured e-liquids is generally higher among young people and young adults than in older adults. It is also more frequent among non-smokers than conventional cigarette smokers *(39)*. The preferences and demand for flavoured nicotine products seem to apply to conventional cigarettes and EN&NNDS interchangeably. Users tend to seek rewards from flavours across the whole range of available nicotine products *(40)*. In other words, when the desired flavour is not available from the desired nicotine product, a proportion of users may seek it from a second-choice nicotine product.

Key messages and conclusions

EN&NNDS are not harmless. Although the consequences for long-term effects on morbidity and mortality have not yet been studied sufficiently, EN&NNDS are not safe for young people, pregnant women and adults who have never smoked. While it is expected that use of EN&NNDS in these groups might increase their health risks, non-pregnant adult smokers who completely switch from combustible tobacco cigarettes to use of unadulterated and appropriately regulated EN&NNDS alone might reduce their health risks. This potential has been recognized by WHO *(41)*, NASEM *(16)* and the CDC *(42)*.

As indicated by WHO *(41)*, the key to any policy on EN&NNDS is to "appropriately regulate these products, so as to minimize consequences that may contribute to the tobacco epidemic and to optimize the potential benefits to public health", as well as "avoiding nicotine initiation among non-smokers and particularly youth while maximizing potential benefits for smokers". To strike such a regulatory balance is challenging in view of the existing scientific evidence and the fact that not all countries will have the required regulatory and surveillance capacity *(43)*. WHO Members States that decide to regulate EN&NNDS may consider the options below to attain the policy objectives set by the Conference of the Parties (COP) of the WHO FCTC, which are to *(44)*:

- prevent the initiation of EN&NNDS by non-smokers and young people, with special attention to vulnerable groups;
- minimize as far as possible potential health risks to EN&NNDS users and protect non-users from exposure to their emissions;
- prevent unproven health claims from being made about EN&NNDS; and
- protect tobacco-control activities from all commercial and other vested interests related to EN&NNDS, including interests of the tobacco industry.

Countries that decide to regulate EN&NNDS should consider:

- being mindful of the unintended consequences of any regulatory measure in swaying the market towards any specific type of EN&NNDS product;
- regulating EN&NNDS that make health claims as medicinal products and therapeutic devices and authorizing their marketing once such claims have been verified scientifically;
- banning or restricting advertising, promotion and sponsorship of EN&NNDS, regulating sales channels (including online sales) and strongly enforcing laws on minimum age of purchase, while recognizing that restricting access to tobacco products for minors and adults to make it difficult to transition to cigarettes when using EN&NNDS is paramount;
- minimizing health risks to EN&NNDS users by standardizing:
  - the manufacture of devices and EN&NNDS components under effective electrical equipment safety regulations, including waste and safe disposal of electrical and electronic equipment;

- the content of e-liquids, to limit the amount of nicotine available per cartridge or bottle and avoid some ingredients, such as carcinogens, mutagens or reprotoxins, those that facilitate inhalation or nicotine uptake and additives such as amino acids, caffeine, colouring agents, essential fatty acids, glucuronolactone, probiotics, taurine, vitamins and mineral nutrients – the existing evidence is insufficient to recommend banning (or not banning) certain flavours that may be attractive to children; and
- the packaging of e-liquids by requiring child-proof containers and labelling ENDS to inform users of the addictive nature of the product;
- minimizing health risks to non-users by outlawing the use of EN&NNDS in all indoor spaces or where smoking is prohibited until it is proven that the second-hand aerosol poses no health risks to bystanders;
- limiting the levels and number of specific flavours allowed in EN&NNDS to reduce initiation by young people; and
- setting surveillance systems to monitor the evolution in patterns of EN&NNDS consumption and detect health or safety incidents involving EN&NNDS – given the current state of knowledge about market dynamics, it is extremely important for countries to start monitoring EN&NNDS products in the market and evaluate the impact of regulation on prices and consumption (this includes surveillance of population patterns of EN&NNDS use by use intensity, type of device, the content of e-liquid and reason for use, and by demographic characteristics and smoking status); as the market is rapidly evolving, adjustments to taxation approaches may be needed over time.

In addition, countries that decide to impose an excise tax should consider:

- adopting the best tax structure that the level of national tax administration, product regulation and tobacco-control policies determine – for example, countries with strong tax administration and strong product regulation may find the option of specific excise taxation advantageous, while those with strong tax administration and weak product regulation may find an ad valorem system is an option;
- setting product characteristics to improve the effectiveness of any taxation structure, regardless of context; and
- collecting tax in the same way as for tobacco products in the country (in most countries, the collection is made at the source – the manufacturing/importing point).

Some types of ENDS help some smokers quit under certain circumstances, but the evidence is insufficient to issue a blanket recommendation to use any type of EN&NNDS as a cessation aid for all smokers.

A final and important caveat regarding any EN&NNDS policy, of whatever nature, is that such a policy would benefit extraordinarily from the simultaneous implementation of a very strong tobacco-control policy to curtail any potential trajectory from EN&NNDS use to smoking.

References[3]

1. Dai H, Leventhal A. Prevalence of e-cigarette use among adults in the United States, 2014–2018. JAMA 2019;322(18):1824–27. doi:10.1001/jama.2019.15331.
2. Special Eurobarometer 458: attitudes of Europeans towards tobacco and electronic cigarettes. Brussels: European Commission, Directorate General for Health and Food Safety; 2017 (https://ec.europa.eu/commfrontoffice/publicopinion/index.cfm/ResultDoc/download/DocumentKy/79003).
3. Vaping in England: evidence update summary February 2019. London: Public Health England; 2019 (https://www.gov.uk/government/publications/vaping-in-england-an-evidence-update-february-2019/vaping-in-england-evidence-update-summary-february-2019#vaping-in-young-people).
4. New Zealand Health Survey: use e-cigarettes once a month. In: Annual update of key results 2017/18. New Zealand Health Survey [website]: Wellington: Ministry of Health; 2019 (https://minhealthnz.shinyapps.io/nz-health-survey-2018-19-annual-data-explorer/_w_01f170d8/#!/explore-indicators).
5. Appendix XI, table 11.2 – adult tobacco survey smokeless tobacco or e-cigarettes. In: WHO report on the global tobacco epidemic 2019 [website]. Geneva: World Health Organization; 2019 (https://www.who.int/tobacco/global_report/en/).
6. Special Eurobarometer 429: attitudes of Europeans towards tobacco and electronic cigarettes. Brussels: European Commission, Directorate General for Health and Food Safety; 2015 (https://ec.europa.eu/commfrontoffice/publicopinion/archives/ebs/ebs_429_en.pdf).
7. Prevalence of e-cigarette use. In: Reid JL, Hammond D, Tariq U, Burkhalter R, Rynard VL, Douglas O. Tobacco use in Canada: patterns and trends, 2019 edition. Waterloo (ON): Propel Centre for Population Health Impact, University of Waterloo; 2019:90–7 (https://uwaterloo.ca/tobacco-use-canada/sites/ca.tobacco-use-canada/files/uploads/files/tobacco_use_in_canada_2019.pdf).
8. Appendix XI, table 11.4 – youth tobacco surveys smokeless tobacco or e-cigarettes. In: WHO report on the global tobacco epidemic 2019 [website]. Geneva: World Health Organization; 2019 (https://www.who.int/tobacco/global_report/en/).
9. Yoong SL, Stockings E, Chai LK, Tzelepis F, Wiggers, Oldmeadow C et al. Prevalence of electronic nicotine delivery systems (ENDS) use among youth globally: a systematic review and meta-analysis of country level data. Aust NZ J Public Health 2018;42(3):303–8. doi:10.1111/1753-6405.12777.
10. Cullen K, Ambrose B, Gentzke A, Apelberg B, Jamal A, King B. Notes from the field: use of electronic cigarettes and any tobacco product among middle and high school students — United States, 2011–2018. MMWR Morb Mortal Wkly Rep. 2018;67(45):1276–77. doi:10.15585/mmwr.mm6745a5.
11. Use of e-cigarettes among young people in Great Britain. London: Action on Smoking and Health; 2019 (https://ash.org.uk/wp-content/uploads/2019/06/ASH-Factsheet-Youth-E-cigarette-Use-2019.pdf).
12. E-cigarette use. In: Reid JL, Hammond D, Tariq U, Burkhalter R, Rynard VL, Douglas O. Tobacco use in Canada: patterns and trends, 2019 edition. Waterloo (ON): Propel Centre for Population Health Impact, University of Waterloo; 2019:89–105 (https://uwaterloo.ca/tobacco-use-canada/sites/ca.tobacco-use-canada/files/uploads/files/tobacco_use_in_canada_2019.pdf).
13. Hammond D, Reid JL, Rynard V, Fong GT, Cummings KM, McNeill A et al. Prevalence of vaping and smoking among adolescents in Canada, England, and the United States: repeat national cross-sectional surveys. BMJ 2019;365:l2219. doi:10.1136/bmj.l2219.
14. Historical NYTS data and documentation. In: Centers for Disease Control and Prevention [website]. Atlanta (GA): Centers for Disease Control and Prevention; 2019 (https://www.cdc.gov/tobacco/data_statistics/surveys/nyts/data/index.html).
15. McNeill A, Brose LS, Calder R, Bauld L, Robson D. Evidence review of e-cigarettes and heated tobacco products 2018. A report commissioned by Public Health England. London: Public Health England; 2018 (https://assets.publishing.service.gov.uk/government/uploads/system/uploads/attachment_data/file/684963/Evidence_review_of_e-cigarettes_and_heated_tobacco_products_2018.pdf).

---

3 All weblinks accessed 13 January 2020.

16. The National Academies of Sciences, Engineering, Medicine. Public health consequences of e-cigarettes. Washington (DC): The National Academies Press; 2018 (https://www.ncbi.nlm.nih.gov/pubmed/29894118).

17. POS5-51: PATH study wave 1 biomarkers of inflammation and oxidative stress among adult e-cigarette and cigarette users [research poster]. In: SNRT 25 Rapid Response Abstracts. San Francisco (CA): Society for Research on Nicotine and Tobacco; 2019:24 (https://cdn.ymaws.com/www.srnt.org/resource/resmgr/SRNT19_Rapid_Abstracts.pdf).

18. Wang J, Olgin J, Nah G, Vittinghof E, Cataldo JK, Pletcher MJ et al. Cigarette and e-cigarette dual use and risk of cardiopulmonary symptoms in the Health eHeart Study. PLoS ONE 2018;13(7):e0198681. doi:10.1371/journal.pone.0198681.

19. Outbreak of lung injury associated with e-cigarette use, or vaping, products. In: Centers for Disease Control and Prevention [website]. Atlanta (GA): Centers for Disease Control and Prevention; 2019 (https://www.cdc.gov/tobacco/basic_information/e-cigarettes/severe-lung-disease.html#latest-outbreak-information).

20. Goniewicz ML, Smith DM, Edwards KC, Blount BC, Caldwell KL, Feng J et al. Comparison of nicotine and toxicant exposure in users of electronic cigarettes and combustible cigarettes. JAMA Netw Open 2018;1(8):e185937. doi:10.1001/jamanetworkopen.2018.5937.

21. Borland R, Murray K, Gravely S, Fong GT, Thompson ME, McNeill A et al. A new classification system for describing concurrent use of nicotine vaping products alongside cigarettes (so-called "dual use"): findings from the ITC-4 Country Smoking and Vaping Wave 1 Survey. Addiction 2019;114(S1):24–34. doi:10.1111/add.14570.

22. Robertson L, Hoek J, Blank M, Richards R, Ling P, Popova L. Dual use of electronic nicotine delivery systems (ENDS) and smoked tobacco: a qualitative analysis. Tob Control 2019;28:13–9. doi: 10.1136/tobaccocontrol-2017-054070.

23. Guidelines for implementation of Article 8: protection from exposure to tobacco smoke. WHO Framework Convention on Tobacco Control. Geneva: World Health Organization; 2007 (https://www.who.int/fctc/guidelines/adopted/article_8/en/).

24. Hajek P, Phillips-Waller A, Przulj D, Pesola F, Myers Smith K, Bisal N et al. A randomized trial of e-cigarettes versus nicotine-replacement therapy. New Engl J Med. 2019;380(7):629–37. doi:10.1056/nejmoa1808779.

25. Peruga A, Eissenberg T. Clinical pharmacology of nicotine in electronic nicotine delivery systems. In: WHO TobReg: report on the scientific basis of tobacco product regulation. Seventh report of a WHO study group. Geneva: World Health Organization; 2019:31–74 (WHO Technical Report Series No. 1015; https://apps.who.int/iris/bitstream/handle/10665/329445/9789241210249-eng.pdf?ua=1).

26. Zhu SH, Sun JY, Bonnevie E, Cummins SE, Gamst A, Yin L et al. Four hundred and sixty brands of e-cigarettes and counting: implications for product regulation. Tob Control 2014;23(Suppl. 3):iii3–9. doi:10.1136/tobaccocontrol-2014-051670.

27. Hsu G, Sun J, Zhu S. Evolution of electronic cigarette brands from 2013–2014 to 2016–2017: analysis of brand websites. J Med Internet Res. 2018;20(3):e80. doi:10.2196/jmir.8550.

28. Krüsemann E, Boesveldt S, de Graaf K, Talhout R. An e-liquid flavor wheel: a shared vocabulary based on systematically reviewing e-liquid flavor classifications in literature. Nicotine Tob Res. 2018;21(10):1310–9. doi:10.1093/ntr/nty101.

29. Krishnan-Sarin SS, O'Malley S, Green BG, Pierce JB, Jordt SE. The science of flavour in tobacco products. In: WHO study group on tobacco product regulation. Report on the scientific basis of tobacco product regulation. Seventh report of a WHO study group. Geneva: World Health Organization; 2019:125–42 (WHO Technical Report Series No. 1015; https://apps.who.int/iris/bitstream/handle/10665/329445/9789241210249-eng.pdf?ua=1).

30. Zare S, Nemati M, Zheng Y. A systematic review of consumer preference for e-cigarette attributes: flavor, nicotine strength, and type. PLoS ONE 2018;13(3):e0194145. doi:10.1371/journal.pone.0194145.

31. Soule EK, Sakuma KK, Palafox S, Pokhrel P, Herzog TA, Thompson N et al. Content analysis of internet marketing strategies used to promote flavored electronic cigarettes. Addict Behav. 2019;91:128–35. doi:10.1016/j.addbeh.2018.11.012.

References

32. Farsalinos KE, Romagna G, Tsiapras D, Kyrzopoulos S, Spyrou A, Voudris V. Impact of flavour variability on electronic cigarette use experience: an internet survey. Int J Environ Res Public Health 2013;10(12):7272–82. doi:10.3390/ijerph10127272.

33. Shiffman S, Sembower MA, Pillitteri JL, Gerlach KK, Gitchell JG. The impact of flavor descriptors on nonsmoking teens' and adult smokers' interest in electronic cigarettes. Nicotine Tob Res. 2015;17(10):1255–62. doi:10.1093/ntr/ntu333.

34. Tackett AP, Lechner WV, Meier E, Grant DM, Driskill LM, Tahirkheli NN et al. Biochemically verified smoking cessation and vaping beliefs among vape store customers. Addiction 2015;110(5):868–74. doi:10.1111/add.12878.

35. Audrain-McGovern J, Strasser AA, Wileyto EP. The impact of flavoring on the rewarding and reinforcing value of e-cigarettes with nicotine among young adult smokers. Drug Alcohol Depend. 2016;166:263–7. doi:10.1016/j.drugalcdep.2016.06.030.

36. Kong G, Morean ME, Cavallo DA, Camenga DR, Krishnan-Sarin S. Reasons for electronic cigarette experimentation and discontinuation among adolescents and young adults. Nicotine Tob Res. 2015;17(7):847–54. doi:10.1093/ntr/ntu257.

37. Krishnan-Sarin S, Morean ME, Camenga DR, Cavallo DA, Kong G. E-cigarette use among high school and middle school adolescents in Connecticut. Nicotine Tobacco Res. 2015;17(7):810–8. doi:10.1093/ntr/ntu243.

38. Morean ME, Butler ER, Bold KW, Kong G, Camenga DR, Dana A et al. Preferring more e-cigarette flavors is associated with e-cigarette use frequency among adolescents but not adults. PLoS ONE 2018;13(1):e0189015. doi:10.1371/journal.pone.0189015.

39. Goldenson NI, Leventhal AM, Simpson KA, Barrington-Trimis JL. A review of the use and appeal of flavored electronic cigarettes. Curr Addict Rep. 2019;6(2):98–113. doi:10.1007/s40429-019-00244-4.

40. Buckell J, Marti J, Sindelar JL. Should flavours be banned in cigarettes and e-cigarettes? Evidence on adult smokers and recent quitters from a discrete choice experiment. Tob Control. 2019;28(2):168–75. doi:10.1136/tobaccocontrol-2017-054165.

41. Provisional agenda item 5.5.2: electronic nicotine delivery systems and electronic non-nicotine delivery systems (ENDS/ENNDS). Report by WHO. In: Conference of the Parties to the WHO Framework Convention on Tobacco Control: seventh session, Delhi, India, 7–12 November 2016. Geneva: World Health Organization; 2016 (Document FCTC/COP/7/11; https://www.who.int/tobacco/communications/statements/eletronic-cigarettes-january-2017/en/).

42. Electronic cigarettes: what's the bottom line? Atlanta (GA): Centers for Disease Control and Prevention; 2019 (https://www.cdc.gov/tobacco/basic_information/e-cigarettes/pdfs/Electronic-Cigarettes-Infographic-508.pdf).

43. Tobacco product regulation: basic handbook. Geneva: World Health Organization; 2018 (https://apps.who.int/iris/handle/10665/274262).

44. Decision: electronic nicotine delivery systems and electronic non-nicotine delivery systems. In: Conference of the Parties to the WHO Framework Convention on Tobacco Control: sixth session, Moscow, Russian Federation,13–18 October 2014. Geneva: World Health Organization; 2014 (document FCTC/COP6(9); https://apps.who.int/gb/fctc/E/E_cop6.htm).

COUNTRY CASE STUDIES ON ELECTRONIC NICOTINE AND NON-NICOTINE DELIVERY SYSTEMS REGULATION, 2019

BRAZIL, CANADA, THE REPUBLIC OF KOREA AND THE UNITED KINGDOM

Abstract

Electronic nicotine and non-nicotine delivery systems (EN&NNDS) are a heterogeneous class of products that use an electrically powered coil to heat and turn a liquid into an aerosol, which is inhaled by the user. Throughout the world, governments use different approaches to regulate EN&NNDS in their marketplace. This brief provides examples of the three more typical regulatory approaches, including banning the sale of EN&NNDS, applying tobacco-control legislation to these products, and creating an elaborate group of specific regulations or recommendations related to the sale, marketing, packaging, product regulation, reporting/notification, taxation and use in workplaces and public places.

Keywords

ELECTRONIC NICOTINE DELIVERY SYSTEMS (ENDS)

ELECTRONIC NON-NICOTINE DELIVERY SYSTEMS (ENNDS)

REGULATION

CASE STUDIES

Acknowledgements

This brief was written by Armando Peruga, Consultant, WHO Regional Office for Europe, with contributions from Sarah Galbraith-Emami, Technical Officer, Marine Perraudin, Technical Officer, WHO headquarters, Kristina Mauer-Stender, Programme Manager, and Elizaveta Lebedeva, Consultant, Tobacco Control Programme, Division of Noncommunicable Diseases and Promoting Health through the Life-course, WHO Regional Office for Europe.

The authors would like to thank Nino Berdzuli, acting Director, Division of Noncommunicable Diseases and Promoting Health through the Life-course, WHO Regional Office for Europe, for her overall leadership and support for the development of this briefing.

The publication was made possible by funding from the Government of the Russian Federation and the Government of Germany.

Introduction

Electronic nicotine delivery systems (ENDS) represent a mixed class of products that use an electrically powered coil to heat and turn a liquid into an aerosol, which is inhaled by the user. The liquid generally is made of propylene glycol, glycerol, or a mix of them; it always includes nicotine and may contain flavours. When the liquid does not contain nicotine, they are referred to as electronic non-nicotine delivery systems (ENNDS). Together, ENDS and ENNDS are referred to as EN&NNDS.

Governments use different approaches to regulate EN&NNDS in their marketplace. Four countries have been chosen to depict those different approaches: Brazil, Canada, the Republic of Korea and the United Kingdom.

Canada and the United Kingdom allow the sale of EN&NNDS and regulate these products with specific norms applicable to them, but both have particular features. Canada has a federal system of government by which the national (or federal) level of government regulates some areas of public life through its executive, legislative or judiciary branches. Subnational levels of government can complement federal regulations related to health or regulate on their own. In the United Kingdom, the Government explicitly promotes that smokers switch from conventional tobacco to using ENDS to quit smoking. The Republic of Korea also allows the sale of EN&NNDS but applies existing regulation for conventional tobacco products to ENDS, although not for ENNDS. Brazil bans de facto the sale of both ENDS and ENNDS.

Three of the four countries have a strong tobacco-control environment, indicated by implementation at the highest level of achievement of at least four of the six[1] MPOWER measures (1). In the following sections, the prevalence of EN&NNDS use and the general regulatory approach towards EN&NNDS is described for each country. Annex 1 and Annex 2 present national or federal regulations by policy domain and the specific subnational regulations in the case of Canada.

Fig. 1 compares the prevalence of daily use of ENDS in a sample of combined smokers and ex-smokers in each of the four countries described in this document.

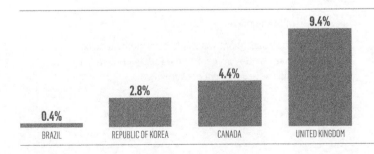

FIG. 1.

Daily use of ENDS among adult (≥ 18 years) smokers and recent ex-smokers in four countries, 2016

Source: modified after Gravely et al. *(2)*

---

1 The six MPOWER measures are: monitor tobacco use and prevention policies, protect people from tobacco smoke, offer help to quit tobacco use, warn about the dangers of tobacco (health warnings), enforce bans on tobacco advertising, promotion and sponsorship, and raise taxes on tobacco.

Brazil

See Box 1 for a summary of EN&NNDS use in Brazil *(2)*.

> **BOX 1.**
> **EN&NNDS USE IN BRAZIL**
>
> The latest data available from Brazil are from 2016 and cover only the prevalence of ENDS use among a sample of 1340 adult smokers and recent ex-smokers from Rio de Janeiro, São Paulo and Porto Alegre. At the time, 1% of interviewees were using ENDS at least once a month and 0.4% at least daily.
>
> *Source:* Gravely et al. *(2)*

Effectively, Brazil bans EN&NNDS. The marketing, importation and advertising of "smoking electronic devices" has been prohibited through the Resolution of the Collegiate Board of the National Health Surveillance Agency of Brazil (ANVISA) (RDC) 46/2009 *(3)*. Smoking electronic devices include EN&NNDS and heated tobacco products (HTPs), but manufacturers seeking to market EN&NNDS in Brazil can submit a request for registration of their product if they have the necessary background information proving the efficacy, effectiveness and safety of the devices. In such cases, ANVISA reviews the information provided and decides whether to register the product, and therefore allow its marketing. To date, no request has been submitted for registration of EN&NNDS. If any is registered, the legislation on tobacco control will apply to EN&NNDS and their use will be banned in all indoor public places, workplaces and public transport.

ANVISA's policy for EN&NNDS continues to be to ban them. At the time, the agency's decision was based on the lack of scientific data on the claims of these products, but ANVISA considers its position regularly by reviewing periodically scientific evidence on the potential health risks and benefits of EN&NNDS (in 2016 *(4)*), technical panels (in 2018 *(5)*) and public hearings (in 2019 *(6)*) with participation from the industry and public health organizations (see ANVISA *(7)* for links to all presentations made in the hearings).

Canada

See Box 2 for a summary of EN&NNDS use in Canada *(8)*.

A federal law, the Tobacco and Vaping Products Act (TVPA) *(9)*, allowed the marketing of ENDS only in 2018: the sale of ENNDS was always permitted. Regulations of ENDS stemming from the TVPA were not wholly developed at federal level when this brief was written. Health Canada nevertheless conducted a public consultation on such regulations *(10)*; the comments received were summarized *(11)* and the Government published the proposed regulations in December 2019. These are now subject to public comment until the end of January 2020 *(12)*. If approved, the regulation will place additional controls to: a) further restrict the promotion of vaping products, including at point of sale and online; b) require health warnings on advertisements; c) prohibit the manufacture of vaping products with certain flavours or flavour ingredients; and d) restrict the concentration and delivery of nicotine in vaping products.

Current federal legislation considers EN&NNDS that make health claims separately from those that do not. If they do, they are governed at federal level by the Food and Drugs Act *(13)* and its regulations, in addition to the TVPA. If EN&NNDS do not make health claims, they are governed by the Canada Consumer Product Safety Act *(14)*, as amended in 2018, and the TVPA of 2018.

The TVPA aims "to prevent vaping product use from leading to the use of tobacco products by young persons and non-users of tobacco products" *(9)*. It establishes a national minimum age of sale for vaping products and significantly restricts their promotion, including bans on lifestyle advertising or promotions that are appealing to young people.

All e-liquids are also subject to the Consumer Chemicals and Containers Regulations of 2001 *(15)*. The Non-smokers' Health Act addresses where to use and where not to use EN&NNDS in federal workplaces *(16)*. Provinces have the authority to expand on the law, including where to use EN&NNDS or not in venues that are not under federal jurisdiction.

EN&NNDS are not treated like tobacco products under Canadian tax legislation. Only the regular general sale tax, which includes a provincial sale tax component, presently applies *(17)*. Two provinces, however, announced in 2019 a tax increase on EN&NNDS for 2020. The Government of British Columbia will increase the provincial sales tax on EN&NNDS from 7% to 20% *(18)*. The tax will apply to all EN&NNDS devices, any substance used in the device, and parts and accessories. The new regulations in British Columbia will also require plain packaging and health warnings for vaping products and restrict public advertising in areas frequented by young people. The Government of Alberta has introduced a fiscal plan *(19)* that includes a 20% tax on the retail sale price of EN&NNDS products in 2020. The tax will

---

BOX 2.
EN&NNDS USE IN CANADA

In 2017:

- 2.9% and 1.0% of Canadians aged 15 and older had used EN&NNDS in the past 30 days or daily, respectively;

- 12.2% of current smokers and 2.4% of non-smokers were past 30-day EN&NNDS users;

- about 64% of adult EN&NNDS users reported that the last time they used a device, the e-liquid contained nicotine, despite nicotine-containing e-liquids not being approved for sale in Canada at that time; and

- fruit and tobacco were the most commonly cited flavours of the last-used EN&NNDS.

Among young people in school grades 7–9 in 2016–2017:

- 5.4% had used an EN&NNDS product in the past 30 days; and

- two thirds of current smokers had used EN&NNDS in the past 30 days, compared to approximately 5% of non-smokers.

Source: Reid et al. *(8)*.

apply to all vaping liquids, including cannabis liquids and do-it-yourself vaping products sold separately for vaping, such as propylene glycol, vegetable, glycerin, nicotine solutions and flavourings, as well as all vaping devices and related accessories *(20)*. In both provinces, the rationale cited for the tax increase is to curtail the rise of EN&NNDS use among young people.

Health Canada actively monitors TVPA compliance among individuals and corporations. It inspected more than 3000 retail establishments in 2019, while tracking online sales and promotion of EN&NNDS and taking action where necessary. Health Canada is also undertaking a public education campaign through advertising, social media and experiential events in schools to increase awareness about the harms and risks associated with vaping product use targeted at young Canadians aged 13–18 *(21,22)*.

Annex 1 contains a description of the federal regulation that applies to vaping products. Canada is a federal country in which some of its subnational jurisdictions, the provinces, have regulated EN&NNDS within their purview. Eight of the 12 provinces and territories have such laws on EN&NNDS (see Annex 2 for a summary of their provisions).

The Republic of Korea

See Box 3 for a summary of EN&NNDS use in the Republic of Korea (2,23,24).

In 2007, the Government amended the scope of the Tobacco Business Act (25) to apply not only to products manufactured using tobacco leaves, but also to those made without using tobacco leaves as the raw materials for inhaling, such as ENDS, except when used as medicines or non-pharmaceutical drugs as covered in the Pharmaceutical Affairs Act (26). Article 27-2 of the enforcement decree (27) of the National Health Promotion Act (28) in 2017 names ENDS specifically as electronic cigarettes and defines them as products "made to cause the same effect as smoking by inhaling nicotine-containing solution or shredded tobacco into the body through respiratory organ with an electronic device."

ENDS and HTPs are classified in the Republic of Korea as "electronic tobacco" products, so most tobacco-control legislation applies to ENDS and HTPs.

Using ENDS is not allowed where smoking is banned. Smoking is unlawful in all parts of indoor health-care and educational facilities, except universities. Smoking is also prohibited in designated non-smoking areas of workplaces and public places.

Advertising of ENDS is illegal on TV, radio and billboards and other outdoor facilities. Some forms of marketing, such as promotional discounts, are also barred. Enforcement has been challenging, however, typified by the refusal of British American Tobacco Korea to take down an ad in the form of a music video on the grounds that it showcased "an electronic device" and not the container that holds the actual nicotine product. At the same time, British American Tobacco offered an aggressive 50% discount for its vaping devices (29). Other incidents include an ENDS company targeting young people through the promotion of a film in online forums for young people with the option of winning free tickets to watch the movie, the placement of EN&NNDS in films for teenagers and an EN&NNDS company sponsoring a community-based youth orchestra (30).

ENDS must carry pictorial health warnings occupying 50% of the main surfaces of the package (31). ENDS packaging and advertisements should include health-warning text that indicates they contain harmful substances such as tobacco-specific nitrosamines and formaldehyde.

ENDS are subject to a tobacco consumption tax of 628 Won (approximately US$ 0.5) per mL of nicotine solution (32). They are also subject to other taxes and charges (national health promotion, local education and individual consumption taxes) in addition to a waste charge of 24 Won per 20 cartridges (approximately US$ 0.02) and 10% value added tax (VAT). In total, ENDS are taxed at 1799 Won per mL (approximately US$ 1.5) of nicotine liquid (33).

> **BOX 3.**
> **EN&NNDS USE IN THE REPUBLIC OF KOREA**
>
> The Korea National Health and Nutrition Examination Survey estimated in 2017 that the prevalence of using EN&NNDS at least once a month among adults of 19+ years was 2.3%. Among young people between 13 and 18 years, the figure was 2.7%, according to the Korea Youth Risk Behaviour Web-based Survey in 2018.
>
> A study among adult current smokers indicated that 5.5% used EN&NNDS at least once a month and 2.8% on a daily basis.
>
> Sources: Gravely (2); WHO (23); WHO (24).

The epidemic of lung disease related to the use of adulterated e-liquids for EN&NNDS in the United States of America in 2019 has prompted the Government of the Republic of Korea to initiate specific actions to curb the use of these products among young people. The Government is acting on four fronts: amendments to legislation, enforcement of existing laws, surveillance and education.

The Ministry of Health and Welfare has submitted to the National Assembly changes to the legislation to close some loopholes in the definition of tobacco classes, which would include novel products and ban some flavours. The specifics of these announced changes to the law are not yet clear. Through this legislation, the Government will also require vaping manufacturers to submit more detailed information on the ingredients and additives of their products.

In enforcing existing laws, the Government is focused on reducing illegal sales of e-cigarettes, both online and to minors. At the same time, the Korean Agency for Technology and Standards is preventing the distribution and sale of illegal batteries for safety reasons and local authorities across the country are carrying out inspections to enforce smoke-free regulations in public places, with a particular focus on the use of ENDS and HTPs *(34)*.

The Government is closely monitoring potential cases of lung diseases that may be related to the use of EN&NNDS through the existing consumer risk-monitoring system. The Ministry of Health and Welfare is conducting a health education campaign aimed at providing information on the health risks associated with EN&NNDS and counselling for smokers on how they can quit.

The United Kingdom

See Box 4 for a summary of EN&NNDS use in the United Kingdom *(35,36)*.

As of December 2019, the United Kingdom was still part of the European Union (EU), so was subject to the European Tobacco Product Directive (TPD) *(37)* that came into effect in May 2016 and was transposed into United Kingdom law via the Tobacco and Related Products Regulations of 2016 *(38)*.

The TPD does not regulate ENDS with medicinal purposes. ENDS and refill containers presented as a remedy to get rid of nicotine addiction or restoring, correcting or modifying physiological functions in a significant manner, or otherwise intended for medical purposes, are subject to Directive 2001/83/EC of the European Parliament and of the Council of 6 November 2001 on the community code relating to medicinal products for human use, Council Directive 93/42/EEC of 14 June 1993 concerning medical devices, and Regulation (EC) 726/2004 of the European Parliament and of the Council of 31 March 2004 *(39)*.

In the United Kingdom, ENDS under these transposed directives and regulations *(40–42)* fall under the purview of the Medicines and Healthcare Products Regulatory Agency (MHRA).

The TPD does not cover ENNDS, which individual EU Member States can regulate. ENNDS are regulated in the United Kingdom under the General Product Safety Regulations 2005 *(43)*.

The regulation of ENDS with no medicinal purposes in the United Kingdom:

- sets safety standards for e-liquid containers, such as being child- and tamper-proof, being protected against breakage and leakage and having a mechanism that ensures refilling without leakage;
- limits the volume of liquid receptacles in disposable ENDS, single-use cartridges and tanks to less than or equal to (≤) 2 mL;
- limits the volume of dedicated refill e-liquid containers to ≤ 10 mL and the maximum concentration of nicotine per amount of e-liquid to ≤ 20 mg/mL;
- bans the following substances in e-liquids:
  - additives that create the impression that a tobacco product has a health benefit or presents reduced health risks, such as vitamins;
  - additives that are associated with energy and vitality, such as caffeine or taurine;
  - additives having colouring properties for emissions; and
  - ingredients that pose a risk to human health in heated or unheated form (except for nicotine);
- requires devices to deliver a consistent dose of nicotine under normal conditions;
- sets labelling requirements such as carrying information on possible adverse effects or addictiveness and toxicity, including a list of ingredients;

> **BOX 4.**
> **EN&NNDS USE IN THE UNITED KINGDOM**
>
> According to the Eurobarometer, 5.6% of adults used EN&NNDS at least once a month in 2017, which represented close to 3 million users. The United Kingdom has the highest prevalence of current use of EN&NNDS in the European Union. Between 18% and 20% of current smokers in 2017 were also EN&NNDS users, while around 9% of ex-smokers and 0.3–0.6% of never-smokers were EN&NNDS users.
>
> Among adults who had ever tried EN&NNDS more than twice in 2017, 42% indicated that they did it to help stop smoking entirely. The proportion of use of EN&NNDS at least weekly among 11–16-year-olds was 1–3% in 2015–2017, depending on the survey. The proportion of never smokers of that age who used EN&NNDS regularly in 2015–2017 was 0.1–0.5%.
>
> *Sources*: House of Commons Science and Technology Committee *(35)*; European Commission *(36)*.

- sets warning requirements about the addictiveness of nicotine;
- requires all ENDS and e-liquids to be notified to MHRA before they can be sold; by the end of 2017, almost 400 producers had submitted information about 32 407 e-liquids (90% of notifications) or devices (10% of notifications);
- allows consumers and health-care professionals to report side-effects and safety concerns with ENDS or refill containers to MHRA through the Yellow Card reporting system;
- bans the sale and provision of EN&NNDS and e-liquids to persons under the age of 18; and
- bans advertising of ENDS devices and e-liquids on TV, radio, the Internet and specific printed publications.

The TPD does not regulate whether to tax ENDS or how. It leaves these decisions to individual Member States. EN&NNDS currently are taxed in the United Kingdom as a consumer product with a 20% VAT rate. There are no specific taxes for EN&NNDS in the United Kingdom. Tobacco products are taxed at a higher rate than EN&NNDS, with specific taxes in addition to VAT. In the third quarter of 2019, a 10 mL nicotine-containing e-liquid was 4.3 times more affordable than a 20-cigarette pack of Marlboro *(44)*.

The TPD also does not regulate where ENDS can and cannot be used (ENDS aerosol-free areas), with Member States once again at liberty to decide. The United Kingdom has not legislated to restrict where EN&NNDS can be used. Wales attempted to limit the use of EN&NNDS in some public places, but the bill was voted down. An undetermined number of workplaces, public places and transportation systems nevertheless have voluntarily banned the use of EN&NNDS where smoking is prohibited.

The context of regulatory efforts in the United Kingdom is characterized by an active scientific and technical debate. Discussion has been influenced by several reports, specifically those commissioned by Public Health England,[2] the latest of which was published in 2019 *(45)*, and those from the Royal College of Physicians in 2016 *(46)*, the British Medical Association in 2018 *(47)* and the House of Commons Science and Technology Committee in 2018 *(35)*. All these reports agree that EN&NNDS are substantially less harmful to health than smoking but are not risk-free. They also indicate that evidence on the long-term health impact is lacking. They signify or imply that ENDS have been beneficial to public health in the United Kingdom and that their promotion as a substitute for smoking is therefore likely to generate significant health gains. Similarly, they indicate or imply that concerns about the risk of EN&NNDS potentially providing a gateway into conventional smoking have not materialized in the United Kingdom *(36,48)*.

Several of these reports describe the lack of high-quality research into the effectiveness of ENDS as a cessation aid. They nevertheless agree that most reported studies demonstrate a positive relationship between ENDS use and smoking cessation. The report from the House of Commons Science

---

[2] Public Health England is an independent executive agency of the Department of Health and Social Care responsible for making the public healthier and reducing differences between the health of different groups by promoting healthier lifestyles, advising government and supporting action by local government, the National Health Service and the public.

and Technology Committee *(35)* recommends that the National Health Service (NHS) in England should set a clear central NHS policy on ENDS in mental health facilities, allowing ENDS use by patients as the default policy unless an NHS trust can show evidence-based reasons for not doing so. The Government responded to the report with a command paper which broadly accepted the Committee's recommendations *(49)*. Regarding ENDS as cessation aids, Public Heath England indicated that *(45)*:

> Combining EC [electronic cigarettes] (the most popular source of support used by smokers in the general population), with stop smoking service support (the most effective type of support), should be a recommended option available to all smokers.

The National Institute for Health and Care Excellence (NICE)[3] recommended in March 2018 that health and social services should explain to people who smoke and who are using, or interested in using, an ENDS product to quit smoking that while these products are not licensed medicines, they are regulated, and many people have found them helpful in quitting smoking cigarettes. NICE also recommended that people using ENDS should stop smoking tobacco entirely because any smoking is harmful. The NHS long-term plan for England recommends a new universal smoking cessation offer for long-term users of specialist mental health and learning disability services, which will include the option for smokers to switch to e-cigarettes while in inpatient settings.

The report from the House of Commons Science and Technology Committee *(35)* expressed concerns about some restrictions established by the TPD. The Committee signalled that norms on the size of tanks and refill containers, the maximum nicotine concentration and advertising were holding back their use as a stop-smoking measure, and that these could be changed following the United Kingdom's departure from the EU. The Committee also considered that *(35)*:

> the level of taxation on smoking-related products should directly correspond to the health risks that they present, to encourage less harmful consumption. Applying that logic, e-cigarettes should remain the least-taxed and conventional cigarettes the most, with heat-not-burn products falling between the two.

Following a consultation, the Committee of Advertising Practice and the Broadcast Committee of Advertising Practice announced that they were lifting the blanket ban on making health claims in non-broadcast advertising for ENDS in media not regulated under the TPD (outdoor advertising, posters on public transport, cinemas, leaflets and direct mail). It currently is unclear how the new guidance will be applied in practice.

---

3 NICE is an independent governmental agency that provides national guidance and advice to improve health and social care.

Conclusions

In 2014, the Conference of the Parties of the WHO Framework Convention on Tobacco Control invited Parties to *(50)*:

> consider prohibiting or regulating ENDS/ENNDS, including as tobacco products, medicinal products, consumer products, or other categories, as appropriate, taking into account a high level of protection for human health.

The Secretariat of the Convention reported in 2018 that 77 Parties regulated or banned EN&NNDS *(51)*. By the end of 2019, this figure was reported as 98 *(52)*.

This brief provides examples of the three more typical regulatory approaches. The first is banning the sale of EN&NNDS. Brazil is a case in point, with the exception that the existing legislation allows the Government to reconsider the decision as soon as a manufacturer presents convincing evidence to support change. The second is the Republic of Korea. This country applies most of its tobacco-control legislation to EN&NNDS, with some exceptions. Finally, Canada and the United Kingdom (the latter as part of the EU) have created an elaborate group of specific regulations or recommendations related to the sale (including minimum age), advertising, promotion, sponsorship, packaging (child-safety packaging, health-warning labelling and trademark), product regulation (nicotine volume/concentration, safety/hygiene, ingredients/ flavours), reporting/notification, taxation, and use in workplaces and public places of EN&NNDS products. This brief has sought to reflect the regulatory approach of the two countries in this category for a range of reasons, which includes the different role of subnational jurisdictions in regulating EN&NNDS and the fact that Canada tends specifically to regulate some aspects of ENNDS while the United Kingdom does not.

It is hoped that these case studies provide information to regulators and public health advocates seeking to explore what kind of EN&NNDS regulatory options are available in practice, so far.

References[4]

1. WHO report on the global tobacco epidemic, 2019. Geneva: World Health Organization; 2019 (https://www.who.int/tobacco/global_report/en/).

2. Gravely S, Driezen P, Ouimet J, Quah ACK, Cummings KM, Thompson ME et al. Prevalence of awareness, ever-use and current use of nicotine vaping products (NVPs) among adult current smokers and ex-smokers in 14 countries with differing regulations on sales and marketing of NVPs: cross-sectional findings from the ITC Project. Addiction 2019;114(6):1060–73. doi:10.1111/add.14558.

3. Resolution of the Collegiate Board – RDC number 46, 28 August 2009. Prohibits the sale, import and advertising of any electronic smoking devices, known as electronic cigarettes. Brasília: Ministry of Health, Brazilian Health Regulatory Agency (ANVISA); 2009 (http://portal.anvisa.gov.br/documents/10181/2718376/RDC_46_2009_COMP.pdf/2148a322-03ad-42c3-b5ba-718243bd1919) (in Portuguese).

4. Electronic cigarettes: what do we know? Study on the composition of vapor and health damage, the role in harm reduction and in the treatment of nicotine dependence. Rio de Janeiro: Ministry of Health, National Cancer Institute José Alencar Gomes da Silva (INCA); 2016 (http://portal.anvisa.gov.br/documents/106510/106594/Livro+Cigarros+eletr%C3%B4nicos+o+que+sabemos/e8a169d0-fd20-4fdc-b11f-ec9281f49700) (in Portuguese).

5. Panel debate: electronic smoking devices [news story] In: ANVISA [website]. Brasília: Brazilian Health Regulatory Agency (ANVISA); 2018 (http://portal.anvisa.gov.br/noticias?p_p_id=101_INSTANCE_FXrpx9qY7FbU&p_p_col_id=column-2&p_p_col_pos=1&p_p_col_count=2&_101_INSTANCE_FXrpx9qY7FbU_groupId=219201&_101_INSTANCE_FXrpx9qY7FbU_urlTitle=painel-debate-dispositivos-eletronicos-para-fumar&_101_INSTANCE_FXrpx9qY7FbU_struts_action=%2Fasset_publisher%2Fview_content&_101_INSTANCE_FXrpx9qY7FbU_assetEntryId=4289520&_101_INSTANCE_FXrpx9qY7FbU_type=content) (in Portuguese).

6. Public hearings 27/08/19 – Process number: 25351.911221/2019-74. Brasília: Brazilian Health Regulatory Agency (ANVISA); 2019 (in Portuguese).

7. Public hearings 08/08/2019 – Process number: 25351.911221/2019-74. Brasília: Brazilian Health Regulatory Agency (ANVISA); 2019 (http://portal.anvisa.gov.br/audiencias-publicas#/visualizar/400068) (in Portuguese).

8. Reid OJ, Hammond D, Tariq U, Burkhalter R, Rynard V, Douglas O. Tobacco use in Canada: patterns and trends, 2019 edition. Waterloo (ON): Propel Centre for Population Health Impact, University of Waterloo; 2019 (https://uwaterloo.ca/tobacco-use-canada/tobacco-use-canada-patterns-and-trends).

9. Tobacco and Vaping Products Act 2017. Ottawa (ON): Government of Canada; last amended 9 November 2019 (https://laws-lois.justice.gc.ca/eng/acts/T-11.5/).

10. Notice of intent – potential measures to reduce the impact of vaping products advertising on youth and non-users of tobacco products. In: Government of Canada [website]. Ottawa (ON): Government of Canada; 2019 (https://www.canada.ca/en/health-canada/programs/consultation-measures-reduce-impact-vaping-products-advertising-youth-non-users-tobacco-products/notice-document.html).

11. Consultation summary: notice of intent – potential measures to reduce the impact of vaping products advertising on youth and non-users of tobacco products. Ottawa: Health Canada; 2019 (https://www.canada.ca/en/health-canada/programs/consultation-measures-reduce-impact-vaping-products-advertising-youth-non-users-tobacco-products/notice-document/summary.html).

12. Vaping Products Promotion Regulations. Canada Gazette Part 1 2019;153(51) (http://www.gazette.gc.ca/rp-pr/p1/2019/2019-12-21/html/reg1-eng.html).

13. Food and Drugs Act 1985. Ottawa (ON): Government of Canada; last amended 21 June 2019 (https://laws-lois.justice.gc.ca/eng/acts/F-27/index.html).

14. Canada Consumer Product Safety Act 2010. Ottawa (ON): Government of Canada; last amended 18 October 2018 (https://laws-lois.justice.gc.ca/eng/acts/C-1.68/).

---

4 All weblinks accessed 18 March 2020.

References

15. Consumer Chemicals and Containers Regulations, 2001. Ottawa (ON): Government of Canada; last amended 22 June 2016 (https://laws-lois.justice.gc.ca/eng/regulations/sor-2001-269/index.html).

16. Non-smokers' Health Act 1985. Ottawa (ON): Government of Canada; last amended 17 October 2018 (https://laws-lois.justice.gc.ca/eng/acts/N-23.6/page-1.html).

17. Walker K. Tobacco taxes not applicable to e-cigarettes. Canadian Tax Focus 2017;7(3):11–12 (https://www.ctf.ca/ctfweb/EN/Newsletters/Canadian_Tax_Focus/2017/3/170315.aspx).

18. Siekierska A. B.C. hikes tax on vaping products from 7% to 20% [news story]. In: Yahoo Finance [website]. Sunnyvale (CA): Yahoo; 2019 (https://ca.finance.yahoo.com/news/bc-hikes-tax-on-vaping-products-205048362.html).

19. Fiscal plan: a plan for jobs and the economy 2019–23. Edmonton (AB): Alberta Treasury Board and Finance; 2019 (https://open.alberta.ca/dataset/3d732c88-68b0-4328-9e52-5d3273527204/resource/2b82a075-f8c2-4586-a2d8-3ce8528a24e1/download/budget-2019-fiscal-plan-2019-23.pdf).

20. Budget 2020: fiscal plan. A plan for jobs and the economy 2020–23. Edmonton (AB): Alberta Treasury Board and Finance; 2020 (https://open.alberta.ca/dataset/05bd4008-c8e3-4c84-949e-cc18170bc7f7/resource/79caa22e-e417-44bd-8cac-64d7bb045509/download/budget-2020-fiscal-plan-2020-23.pdf).

21. Consider the consequences of vaping. In: Government of Canada [website]. Ottawa (ON): Government of Canada; 2019 (https://www.canada.ca/en/services/health/campaigns/vaping.html).

22. Consider the consequences of vaping (health information video). In: Government of Canada [website]. Ottawa (ON): Government of Canada; 2019 (https://youtu.be/mGaDhpXHWrQ).

23. Appendix XI, Table 11.2. Adult tobacco survey smokeless tobacco or e-cigarettes. In: WHO report on the global tobacco epidemic, 2019. Geneva: World Health Organization; 2019 (https://www.who.int/tobacco/global_report/en/).

24. Appendix XI, Table 11.4. Youth tobacco surveys smokeless tobacco or e-cigarettes. In: WHO report on the global tobacco epidemic, 2019. Geneva: World Health Organization; 2019 (https://www.who.int/tobacco/global_report/en/).

25. Tobacco Business Act 1988. Seoul: National Assembly of the Republic of Korea; last amended 26 July 2017 (https://elaw.klri.re.kr/eng_service/lawView.do?hseq=45814&lang=ENG).

26. Pharmaceutical Affairs Act 2007. Seoul: National Assembly of the Republic of Korea; last amended 2 December 2016 (https://elaw.klri.re.kr/eng_service/lawView.do?hseq=40196&lang=ENG).

27. Enforcement Decree of the National Health Promotion Act. Presidential Decree No. 28071, May 29, 2017. Seoul: National Assembly of the Republic of Korea; 2017 (https://elaw.klri.re.kr/kor_service/lawView.do?hseq=43548&lang=ENG).

28. National Health Promotion Act 2017. Seoul: National Assembly of the Republic of Korea; last amended 30 December 2017 (https://elaw.klri.re.kr/kor_service/lawView.do?lang=ENG&hseq=48657).

29. Lee S. Health ministry moves to regulate e-cigarette ads [news story]. In: The Korea Times [website]. Seoul: The Korea Times Co.; 2019 (http://www.koreatimes.co.kr/www/nation/2019/09/119_275619.html).

30. Lee WB. E-cigarette marketing targeted to youth in South Korea. Tob Control 2017;26(e2):e140–4. doi:10.1136/tobaccocontrol-2016-053448.

31. Cigarette warning picture and phrase replacement: all e-cigarettes with pictures symbolizing "carcinogenicity". Seoul: Ministry of Health and Welfare; 2018 (http://www.mohw.go.kr/react/al/sal0301vw.jsp?PAR_MENU_ID=04&MENU_ID=0403&page=4&CONT_SEQ=344802) (in Korean).

32. Enforcement Decree of the Local Tax Act. Presidential Decree No. 28714, 27 March, 2018. Seoul: National Assembly of the Republic of Korea; 2018 (https://elaw.klri.re.kr/eng_service/lawView.do?hseq=47411&lang=ENG).

33. Reducing tobacco use through taxation: the experience of the Republic of Korea. Washington (DC): World Bank Group; 2018 (http://documents.worldbank.org/curated/en/150681529071812689/pdf/127248-WP-PUBLIC-ADD-SERIES-WBGTobaccoKoreaFinalweb.pdf).

34. Intensive crackdown on smoking in the non-smoking area. Seoul: Ministry of Health and Welfare; 2019 (http://www.mohw.go.kr/react/al/sal0301vw.jsp?PAR_MENU_ID=04&MENU_ID=0403&page=1&CONT_SEQ=350874) (in Korean).

35. E-cigarettes. Seventh report of session 2017–19. London: House of Commons Science and Technology Committee; 2018 (https://publications.parliament.uk/pa/cm201719/cmselect/cmsctech/505/505.pdf).

36. Attitudes of Europeans towards tobacco and electronic cigarettes. Special Eurobarometer 458. Brussels: European Commission; 2017 (https://ec.europa.eu/commfrontoffice/publicopinion/index.cfm/ResultDoc/download/DocumentKy/79003).

37. Directive 2014/40/EU of the European Parliament and of the Council of 3 April 2014 on the approximation of the laws, regulations and administrative provisions of the member states concerning the manufacture, presentation and sale of tobacco and related products and repealing Directive 2001/37/EC text with EEA relevance. Brussels: European Union; 2014 (https://eur-lex.europa.eu/legal-content/EN/TXT/?uri=OJ%3AJOL_2014_127_R_0001).

38. The Tobacco and Related Products Regulations 2016. Statutory Instruments 2016 No. 507. London: The Stationery Office; 2016 (http://www.legislation.gov.uk/uksi/2016/507/contents/made).

39. Legal framework governing medicinal products for human use in the EU. In: European Commission [website]. Brussels: European Commission; 2019 (https://ec.europa.eu/health/human-use/legal-framework_en).

40. The Medicines (Codification Amendments Etc.) Regulations 2002. Statutory Instruments 2002 No. 236. London: HMSO; 2002 (http://www.legislation.gov.uk/uksi/2002/236/pdfs/uksi_20020236_en.pdf).

41. The Medicines for Human Use (Fees Amendments) Regulations 2006. Statutory Instruments 2006 No. 494. London: HMSO; 2006 (https://www.legislation.gov.uk/uksi/2006/494/contents/made).

42. The Medicines for Human Use (National Rules for Homeopathic Products) Regulations 2006. Statutory Instruments 2006 No. 1952. London: HMSO; 2006 (http://www.legislation.gov.uk/uksi/2006/1952/contents/made).

43. The General Product Safety Regulations 2005. Statutory Instruments 2005 No. 1803. London: HMSO; 2005 (http://www.legislation.gov.uk/uksi/2005/1803/pdfs/uksi_20051803_en.pdf).

44. Anastasopoulou S. Overview of the EU electronic cigarette market. Tabexpo Congress 2019, 11–14 November, Amsterdam, the Netherlands [conference presentation]. In: ECigIntelligence [website]. London: Tamarind Media Limited; 2019 (https://ecigintelligence.com/wp-content/uploads/2019/11/TABEXPO-2019_Amsterdam_Stavroula_Anastasopoulou_ECigIntelligence.pdf).

45. McNeill A, Brose LS, Calder R, Bauld L. Vaping in England: an evidence update February 2019. A report commissioned by Public Health England. London: Public Health England Publications; 2019 (https://assets.publishing.service.gov.uk/government/uploads/system/uploads/attachment_data/file/821179/Vaping_in_England_an_evidence_update_February_2019.pdf).

46. Nicotine without smoke: tobacco harm reduction. A report by the Tobacco Advisory Group of the Royal College of Physicians. London: Royal College of Physicians; 2016 (https://www.rcplondon.ac.uk/file/3563/download).

47. E-cigarettes: balancing risks and opportunities. London: British Medical Association; 2019 (https://www.bma.org.uk/collective-voice/policy-and-research/public-and-population-health/tobacco/e-cigarettes).

48. McNeill A, Brose L, Calder R, Bauld L, Robson D. Evidence review of e-cigarettes and heated tobacco products 2018. A report commissioned by Public Health England. London: Public Health England Publications; 2018 (https://assets.publishing.service.gov.uk/government/uploads/system/uploads/attachment_data/file/684963/Evidence_review_of_e-cigarettes_and_heated_tobacco_products_2018.pdf).

49. Secretary of State for Health and Social Care. The Government response to the Science and Technology Committee's seventh report of the Session 2017–19 on e-cigarettes. London: The Stationery Office; 2018 (https://assets.publishing.service.gov.uk/government/uploads/system/uploads/attachment_data/file/762847/government-response-to-science-and-technology-committee_s-report-on-e-cig.pdf).

50. Decision: electronic nicotine delivery systems and electronic non-nicotine delivery systems. In: Conference of the Parties to the WHO Framework Convention on Tobacco Control: sixth

session, Moscow, Russian Federation, 13–18 October 2014. Geneva: World Health Organization; 2014 (document FCTC/COP6(9); https://apps.who.int/gb/fctc/E/E_cop6.htm).

51. Report: progress report on regulatory and market developments on electronic nicotine delivery systems (ENDS) and electronic non-nicotine delivery systems (ENNDS). In: Conference of the Parties to the WHO Framework Convention on Tobacco Control: eighth session, Geneva, Switzerland, 1–6 October 2018. Geneva: World Health Organization; 2014 (document FCTC/COP/8/10; https://www.who.int/fctc/cop/sessions/cop8/FCTC_COP_8_10-EN.pdf?ua=1).

52. Country laws regulating e-cigarettes. In: Global Tobacco Control [website]. Baltimore (MD): Global Tobacco Control; 2020 (https://www.globaltobaccocontrol.org/e-cigarette_policyscan).

Annex 1 National or federal regulation that applies to EN&NNDS

This annex shows specific regulation of electronic nicotine and non-nicotine delivery systems (EN&NNDS) by country and policy domain.

Policy domain	Brazil	Canada
Product classifi	EN&NNDS are referred to in the legislation *(1)* as "smoking electronic devices". They are implicitly classifi as tobacco products. "Smoking electronic devices" also include heated tobacco products.	The Tobacco and Vaping Products Act *(2)* defines tobacco products as made in whole or in part of tobacco, including tobacco leaves, as well as devices necessary for the use of such products (such as tobacco-heating devices). However, it defi vaping products very loosely as devices emitting an aerosol for human inhalation and the substances intended for use with those devices, with or without nicotine. Vaping products therefore include zero-nicotine e-liquids.
Pre-marketing notifi to government	Pre-marketing notification to the government is not required.	Pre-marketing notification to the government is not required.
Government pre-marketing approval	Government pre-marketing approval (registration) is mandatory. The National Health Surveillance Agency of Brazil (ANVISA) may authorize registration for the marketing of any "smoking electronic devices" based on the submission of toxicological studies and specific scientific tests to substantiate their efficacy, effectiveness and safety.	Government pre-marketing approval is required if a vaping product is marketed for a therapeutic purpose.
Import, sale and distribution	The import, sale and distribution of EN&NNDS are banned unless ANVISA has previously registered the product. Presently, no manufacturer has submitted any product for registration, so none has been registered. Even if registered, the sale, supply (even free of charge) and distribution of any electronic smoking devices to minors is prohibited.	The sale and supply of EN&NNDS (vaping products in the legislation) to persons under 18 or 19 years of age, depending on the province, are banned. One province, Nova Scotia, also bans the possession of vaping products by minors. The manufacture, importation *(9)* or sale of e-liquids with ≥ 66 mg/mL of nicotine is prohibited under section 38 of the Consumer Chemicals and Containers Regulations (CCCR), 2001 *(10)*.

Republic of Korea	United Kingdom
ENNDS are considered consumer products, while ENDS are classified as tobacco products under articles 2 and 3 of the Tobacco Business Act *(3)*. Article 27-2 of the enforcement decree of the National Health Promotion Act *(4)* defines ENDS as products "made to cause the same effect as smoking by inhaling nicotine-contained solution or shredded tobacco into the body through respiratory organ with an electronic device."	Under the European Tobacco Product Directive (TPD) *(5)* regulatory framework, EN&NNDS and ENDS products that do not make any health claim are classified as consumer products, while EN&NNDS that make health claims are considered medicinal products. Although one ENDS product has been licensed as a medicine *(6)*, it is not currently available on the market.
Pre-marketing notification to the government is not required.	Producers of all devices and e-liquids that were on the market before May 2016 had until November 2016 to submit a notification to the Medicines and Healthcare Products Regulatory Agency (MHRA) *(7)*. Producers of new devices and e-liquid products must submit a notification six months before they intend to put their product on the market. Pre-marketing information must notify toxicological data regarding ingredients (including in heated form) and emissions with potential health impact, including addictiveness *(8)*. An "ingredient" is any substance or element present in a finished product or related product, including paper, filter, ink, capsules, adhesives and any additive.
Government pre-marketing approval is not required.	Government pre-marketing approval is only required for EN&NNDS with a therapeutic purpose.
Sale of ENDS is prohibited to minors (under 19 years). Such a ban does not apply to ENNDS.	Purchase of "nicotine inhaling products" (ENDS and conventional cigarettes) by or for persons under the age of 18 has been banned since 1 October 2015 in England and Wales and since 1 April 2017 in Scotland *(11)*. Such a ban does not apply to ENNDS. Retailers in the European Economic Area (EEA) or a Third Country must complete the registration process as required by the TPD before making sales of tobacco or ENDS (or both) into the United Kingdom. Retailers from the United Kingdom only need to register if they are planning to sell directly to consumers in another EEA state.

Policy domain	Brazil	Canada
Advertising, promotion and sponsorship	The advertising, promotion and sponsorship of EN&NNDS are banned unless ANVISA has previously registered the product. If any are registered, it is reasonable to expect that the same regulation will apply than for smoking products, whose advertising and promotion is prohibited, with a sole exemption granted for the display of the products at the point of sale. There are some restrictions on tobacco sponsorship.	The advertising of e-liquids with ≥ 66 mg/mL of nicotine is prohibited under section 38 of the CCCR, 2001 *(10)*. Otherwise, advertising and promotion of vaping products are banned if: • they are appealing to persons under 18 years of age or the product has an appearance or a function that could make the product appealing to said persons; • they are using lifestyle advertising, testimonials or endorsements (including the depiction of a person, character or animal, whether real or fictional), however displayed or communicated, including through the packaging; • it is likely to create an association between the brand element or the name and a person, entity, event, activity or permanent facility (sponsorship) or uses, directly or indirectly, a vaping product-related brand element or the name of a vaping product manufacturer in the promotional material related to a person, entity, event, activity or permanent facility; • they are displaying a vaping product-related brand element or the name of a vaping product manufacturer on a permanent facility, as part of the name of the facility or otherwise, if the facility is used for a sports or cultural event or activity; • they are presented in a manner that is false, misleading or deceptive; • it could cause a person to believe that health benefits may be derived from the use of the product or its emissions; and • it could discourage tobacco cessation or encourage the resumed use of tobacco products.

Continued

Republic of Korea	United Kingdom
Advertising of ENDS is illegal on TV, radio and billboards and other outdoor supports. Some forms of marketing, such as promotional discounts, are also barred.	ENNDS are not explicitly regulated. In this case, advertising and promotion are governed by the Consumer Protection from Unfair Trading Regulations, which protect consumers from deception or harassment (12). Advertising and promotion of ENDS with health claims to the general public, if approved by the MHRA, could only be legal if considered over-the-counter medicines (13). The following requirements are only applicable to e-liquids, disposable devices and cartridges containing nicotine that are not medicinal products: • cross-border advertising of ENDS and advertising in broadcast TV and radio is prohibited; and • advertising of ENDS is also banned in newspapers, magazines and periodicals, commercial classified ads, commercial email and text messaging (unless explicitly opted in), marketers' online activities (except factual information), promotional marketing online and online (display) ads in paid-for space, paid-for search listings, preferential listings on price-comparison sites, viral advertisements, paid social media placements, advertisement features and contextually targeted branded content, in-game and in-app advertising, and advertisements that are pushed electronically to devices or distributed through web widgets, affiliate links and product placement. E-liquids, disposable devices and cartridges containing nicotine that are not medicinal products are allowed in: • outdoor advertising, including digital outdoor advertising; • posters on public transport (not leaving the United Kingdom); • cinema, direct mail and leaflets; • private, bespoke correspondence between a marketer and a consumer; - media targeted exclusively to the trade; • advertisements for businesses in non-broadcast media; and • sponsorship of events that are not across borders. Any advertising must: • ensure the ads are socially responsible; • not target, feature or appeal to children; • not confuse e-cigarettes with tobacco products; • not make medicinal claims and take care with health claims; and • not mislead about product ingredients or where they may be used. Public Health England believes that "stating that vaping is at least 95% less harmful than smoking remains a good way to communicate the large difference in relative risk unambiguously so that more smokers are encouraged to make the switch from smoking to vaping" (14).

Policy domain	Brazil	Canada
Packaging and labelling	Packaging Specific legislation or regulation of the packaging of EN&NNDS does not exist at present.	Packaging Canada is considering Vaping Products Labelling and Packaging Regulations *(15)*. Until the proposed regulations are approved and come into force, the following requirements apply: • e-liquids with ≥ 66 mg/mL of nicotine meet the classification of "very toxic" under the CCCR, 2001 and are prohibited from manufacture, import, advertising or sale under section 38 of the CCCR, 2001; and • e-liquids with 10–65 mg/mL of nicotine meet the classification of "toxic" and are subject to all applicable requirements under the CCCR, 2001 for toxic chemicals; stand-alone containers of vaping substances intended for sale at retail are required to be sold in child-resistant containers and to be labelled per the applicable CCCR, 2001 requirements, including a toxic-hazard symbol on the container's main display panel. Health Canada considers that e-liquids with 0.1–9 mg/mL of nicotine are potentially toxic when ingested so therefore must adhere to all requirements of the CCCR, 2001 for "toxic" products, including the requirements for a child-resistant container.
	Labelling requirements Specific legislation or regulation of the labelling of EN&NNDS does not exist at present. However, if any EN&NNDS is registered, it is reasonable to assume that the tobacco product regulation for labelling would apply, including mandatory pictorial health warning occupying 50% of the main surfaces of the packaging.	Labelling requirements Vaping products must carry the following warning: "Nicotine is highly addictive" *(16)*.

Annex 1　National or federal regulation that applies to EN&NNDS

Continued

Republic of Korea	United Kingdom
Packaging No specific legal norm applies to the regulation of packaging of EN&NNDS so far.	Packaging All nicotine-containing receptacles (disposable devices, cartridges and e-liquids boles) must have: child-resistant and tamper-evident packaging; protection against breakage and leakage; and a mechanism for ensuring refilling without leakage. The nicotine-containing liquid must be in: a dedicated refill container not exceeding a volume of 10 mL; a disposable electronic cigarette; or a single-use cartridge or tank that does not exceed a volume of 2 mL.
Labelling requirements ENDS must carry pictorial health warnings occupying 50% of the main surfaces of the package.	Labelling requirements The pack must have: • a health warning: "This product contains nicotine which is a highly addictive substance" ; the text must be prominent in black on a white background covering 30% of the area on the front and back of the unit packet and any container pack; • a list of ingredients in the liquid where they are used in quantities of 0.1% or more; • nicotine content and delivery per dose; • batch number; and • recommendation to keep the product out of reach of children. The accompanying leaflet (unless included on the pack) must have instructions for use and storage, including instructions for refilling where appropriate. The MHRA advises that the information should include appropriate advice on product storage, particularly on how to ensure the battery does not malfunction, contraindications, warnings for specific risk groups and possible adverse effects, addictiveness and toxicity, and contact details of the producer, including a contact within the European Union (EU).

Policy domain	Brazil	Canada
Product regulation of contents and emissions	No specific legal norm applies to the regulation of the contents and emissions of EN&NNDS so far, but no EN&NNDS have yet been registered for marketing. All additives that enhance the flavour and taste of tobacco products to make them more attractive are banned *(17)*. It is likely that such a ban would apply to EN&NNDS if any such products are registered. All characterizing flavours are banned for tobacco products *(17)*. It is likely that such a ban would apply to EN&NNDS, if any such products are registered.	Manufacturers of vaping devices are strongly encouraged (but not legally required) to certify their products to standard ANSI/CAN/UL 8139 on Electrical Systems of Electronic Cigarettes and Vaping Devices and to use lithium-ion batteries that meet standard CAN/CSA-E62133 or equivalent. Chargers provided with the product should be certified to the applicable Canadian national standard by a certification body accredited by the Standards Council of Canada. The diluents used in vaping liquids should be within the specifications of an accepted pharmacopoeia, so solvents of known human toxicity, such as ethylene glycol or diethylene glycol, should not be used. "Ingredient" means any substance used in the manufacture of a tobacco product, vaping product or their components, including any substance used in the manufacture of that substance, and, in respect of a tobacco product, also includes tobacco leaves. Additives prohibited are: amino acids, caffeine, colouring agents, essential fatty acids glucuronolactone, probiotics, taurine, vitamins and mineral nutrients. Emissions are not explicitly regulated. However, Health Canada recommends that the generation of harmful emissions due to thermal decomposition of e-liquids should be as low as reasonably achievable. It is recommended, but not legally required, that all flavourings added to vaping liquids should be of food-grade or higher purity and that substances with known inhalation risks (such as diacetyl and 2,3-pentanedione) should not be used in flavourings. The following flavours are legally banned: confectionery, dessert, cannabis, soft drink and energy drink.
Taxation	No specific legal norm applies to the regulation of the taxation of EN&NNDS so far, but no EN&NNDS have yet been registered for marketing.	Only the regular general sale tax applies to EN&NNDS.

Annex 1 National or federal regulation that applies to EN&NNDS

Continued

Republic of Korea	United Kingdom
No specific legal norm applies to the regulation of the contents and emissions of EN&NNDS so far. Additives are not explicitly regulated. Flavourings can be used but not advertised according to article 9-3 of the National Health Promotion Act (4).	E-liquids without nicotine, whether in disposable devices or separate containers, are regulated under General Product Safety Regulations (18). A product is considered safe if it conforms to: a) a specific health and safety rule of part of the United Kingdom, in the absence of a United Kingdom law; b) a voluntary national standard of the United Kingdom giving effect to an official European standard; or c) otherwise conforms to other specified standards or recommendations, including product safety codes of good practice in the sector concerned, and reasonable consumer expectations concerning safety. The assessment of safety is made regarding the risks and categories of risk covered by the specified requirements of conformity. Nicotine e-liquid ingredients: • must be of high purity • may not pose a risk to human health in a heated or unheated form. The maximum nicotine concentration in e-liquid is 20 mg/mL. Additives prohibited are: vitamins or other additives that create the impression that they have a health benefit or present reduced health risks; caffeine, taurine or other additives and stimulant compounds that are associated with energy and vitality; and additives that have colouring effects on emissions. All other additives are restricted to quantities that do not increase, to a significant or measurable degree, the toxicity, addictiveness or carcinogenic, mutagenic or reprotoxic properties of the product when it is consumed. Emissions are not specifically regulated. Flavours are not specifically regulated.
ENDS are subject to several taxes and charges (national health promotion, tobacco consumption, local education and individual consumption taxes) proportional to 1 799 Won/mL (≈ US$ 1.5) nicotine liquid; also there is a waste charge of 24 won/20 cartridges (≈ US$ 0.02) and a 10% value added tax (VAT) (19).	EN&NNDS are taxed with a VAT rate of 20%.

Policy domain	Brazil	Canada
Use in indoor places	The use of EN&NNDS is banned in all enclosed common areas, including aircraft and public transportation. An enclosed common area is defined as a public or private place, accessible to the general public or for common use, totally or partially enclosed on any of its sides by a wall, partition, roof, awning or covering, whether of a permanent or temporary nature (20).	The Non-smokers' Health Act (21) bans indoor use of EN&NNDS in federally regulated workplaces, such as banks, ferries, passenger aircraft and federal government offices. Most provinces have banned their use where tobacco use is banned.
Protection from commercial interests	The protection of public health from commercial interest from the EN&NNDS industry is not explicitly regulated.	The protection of public health from commercial interest from the EN&NNDS industry is not explicitly regulated.
Surveillance and monitoring	No specific legal norm applies to the regulation of the surveillance and monitoring of EN&NNDS so far, but no EN&NNDS have yet been registered for marketing. In case of irregularities, such as the commercialization, advertising and illegal importation of these products, complaints may be lodged through ANVISA's service channels.	Under the Canada Consumer Product Safety Act (CCPSA), sellers, distributors, importers, manufacturers and suppliers in general of vaping products for commercial purposes must report any health or safety incidents involving one such product to Health Canada and the supplier of said product (24). Depending on the type, the incident must be reported within two or 10 days after the day on which those required to report become aware of the incident. Any person who manufactures, imports, advertises, sells or tests a vaping product for commercial purposes must prepare and maintain documents indicating whom they obtained the product from or to whom they sold it, among other data. They are required to submit records upon request of the Government. The purpose of this requirement is to help improve the traceability of noncompliant products through the supply chain if dangers must be addressed (25). Under the CCPSA, Health Canada has the power to order recalls and other measures and order tests or studies on a product. Enforcement actions taken by Health Canada on noncompliant products depend on the degree of risk associated with noncompliance.

Annex 1   National or federal regulation that applies to EN&NNDS

Continued

Republic of Korea	United Kingdom
	There is no blanket legal requirement regulating the use of ENDS or ENNDS in workplaces and public spaces. The owner or manager of each venue may decide to apply restrictions on a private basis. Although there are no official statistics, it seems that most hospitals and transportation have banned vaping indoors.
The use of ENDS, but not ENNDS, is banned where smoking is banned, and is completely banned in health-care facilities and educational facilities except universities. In all other public places and workplaces, smoking and ENDS use is banned only in designated no-smoking areas.	Vaping on a plane is strictly not allowed by the companies as per International Air Transport Association recommendations (22). Public Health England issued guidance on the use of EN&NNDS in public places and workplaces that emphasizes the distinction between smoking and vaping. The guidance states that "a more enabling approach to vaping may be appropriate to make it an easier choice than smoking" (23). It also recommends that policies be based on evidence of harm to bystanders and risk assessments be informed by evidence. The guidance indicates that the risk to the health of bystanders from second-hand aerosol is extremely low and insufficient to justify prohibiting EN&NNDS use indoors based on international peer-reviewed evidence.
The protection of public health from commercial interest from the EN&NNDS industry is not explicitly regulated.	The protection of public health from commercial interest from the EN&NNDS industry is not explicitly regulated.
No legal norm applies to the regulation of the surveillance and monitoring of EN&NNDS specifically, but the Government is closely monitoring cases of lung diseases potentially related to the use of EN&NNDS through the existing consumer risk-monitoring system (26).	Consumers and health-care professionals can report both side-effects and product safety concerns to the MHRA through the Yellow Card scheme. This scheme records suspected adverse reactions to medicines from health professionals, manufacturers or members of the public. It includes ENDS and e-liquids. A total of 37 reports were received with a suspected adverse reaction to electronic cigarettes between 1 January 2015 and 20 October 2017, and 263 reports associated with a suspected adverse drug reaction to nicotine replacement therapy were received during the same reporting period. The most commonly reported adverse reaction related to gastrointestinal disturbance and respiratory problems.

Reference[5]

1. Resolution of the Collegiate Board – RDC number 46, 28 August 2009. Prohibits the sale, import and advertising of any electronic smoking devices, known as electronic cigarettes. Brasília: Ministry of Health, Brazilian Health Regulatory Agency (ANVISA); 2009 (http://portal.anvisa.gov.br/documents/10181/2718376/RDC_46_2009_COMP.pdf/2148a322-03ad-42c3-b5ba-718243bd1919) (in Portuguese).

2. Tobacco and Vaping Products Act 2017. Ottawa (ON): Government of Canada; last amended 9 November 2019 (https://laws-lois.justice.gc.ca/eng/acts/T-11.5/).

3. Tobacco Business Act 1988. Seoul: National Assembly of the Republic of Korea; last amended 26 July 2017 (https://elaw.klri.re.kr/eng_service/lawView.do?hseq=45814&lang=ENG).

4. National Health Promotion Act 2017. Seoul: National Assembly of the Republic of Korea; last amended 30 December 2017 (https://elaw.klri.re.kr/kor_service/lawView.do?lang=ENG&hseq=48657).

5. Directive 2014/40/EU of the European Parliament and of the Council of 3 April 2014 on the approximation of the laws, regulations and administrative provisions of the member states concerning the manufacture, presentation and sale of tobacco and related products and repealing Directive 2001/37/EC text with EEA relevance. Brussels: European Union; 2014 (https://eur-lex.europa.eu/legal-content/EN/TXT/?uri=OJ%3AJOL_2014_127_R_0001).

6. e-Voke 10mg & 15Mg electronic inhaler. PL 42601/0003-4. London: Medicines & Healthcare Products Regulatory Agency; 2019 (https://mhraproductsprod.blob.core.windows.net/docs-20200224/56f25daab2a2968139bc37075e194d1a5f12b33f).

7. Medicines and Healthcare Products Regulatory Agency. In: GOV.UK [website]. London: Government Digital Service (GDS); 2020 (https://www.gov.uk/government/organisations/medicines-and-healthcare-products-regulatory-agency).

8. The Tobacco and Related Products Regulations 2016. Statutory Instruments 2016 No. 507. London: The Stationery Office; 2016 (https://assets.publishing.service.gov.uk/government/uploads/system/uploads/attachment_data/file/440989/SI_tobacco_products_acc.pdf).

9. Customs Notice 18-05 – Importation of Vaping Products – under the Tobacco and Vaping Products Act (TPVA). In: Canada Border Services Agency [website]. Ottawa (ON): Canada Border Services Agency; 2018 (https://www.cbsa-asfc.gc.ca/publications/cn-ad/cn18-05-eng.html).

10. Consumer Chemicals and Containers Regulations, 2001. Ottawa (ON): Government of Canada; last amended 22 June 2016 (https://laws-lois.justice.gc.ca/eng/regulations/sor-2001-269/index.html).

11. The Nicotine Inhaling Products (Age of Sale and Proxy Purchasing) Regulations 2015. Statutory Instruments 2015 No. 895. London: The Stationery Office; 2015 (http://www.legislation.gov.uk/uksi/2015/895/pdfs/uksi_20150895_en.pdf).

12. Marketing and advertising: the law. In: GOV.UK [website]. London: Government Digital Service (GDS); 2019 (https://www.gov.uk/marketing-advertising-law/regulations-that-affect-advertising).

13. Advertise your medicines: how to comply with the requirements on promoting medicines to the public and to prescribers and suppliers of medicines. In: GOV.UK [website]. London: Government Digital Service (GDS); 2020 (https://www.gov.uk/guidance/advertise-your-medicines#advertise-to-the-public).

14. McNeill A, Brose L, Calder R, Bauld L, Robson D. Evidence review of e-cigarettes and heated tobacco products 2018. A report commissioned by Public Health England. London: Public Health England Publications; 2018 (https://assets.publishing.service.gov.uk/government/uploads/system/uploads/attachment_data/file/684963/Evidence_review_of_e-cigarettes_and_heated_tobacco_products_2018.pdf).

---

5 All weblinks accessed 18 March 2020.

15. Vaping Products Labelling and Packaging Regulations. Canada Gazette Part 1 2019;153(25) (http://gazette.gc.ca/rp-pr/p1/2019/2019-06-22/html/reg4-eng.html).

16. List of health warnings for vaping products. In: Government of Canada [website]. Ottawa (ON): Government of Canada; 2019 (https://www.canada.ca/en/health-canada/services/smoking-tobacco/vaping/product-safety-regulation/list-health-warnings-vaping-products.html).

17. The General Product Safety Regulations 2005. Statutory Instruments 2005 No. 1803. London; HMSO: 2005 (http://www.legislation.gov.uk/uksi/2005/1803/pdfs/uksi_20051803_en.pdf).

18. Resolution of the Collegiate Board – RDC number 14, 15 March 2012. Provides for the maximum limits of tar, nicotine and carbon monoxide in cigarettes and the restriction of the use of additives in tobacco products derived from tobacco, and other measures. Brasília: Ministry of Health, Brazilian Health Regulatory Agency (ANVISA); 2012 (http://bvsms.saude.gov.br/bvs/saudelegis/anvisa/2012/rdc0014_15_03_2012.pdf) (in Portuguese).

19. Reducing tobacco use through taxation: the experience of the Republic of Korea. Washington (DC): World Bank Group; 2018 (http://documents.worldbank.org/curated/en/150681529071812689/pdf/127248-WP-PUBLIC-ADD-SERIES-WBGTobaccoKoreaFinalweb.pdf).

20. Decree number 8.262, of 31 May 2014. Amends Decree number 2.018, of 1 October 1996, which regulates Law number 9.294, of 15 July 1996. Brasília: Legal Affairs Subsection, Presidency of the Republic Civil House; 2014 (http://www.planalto.gov.br/ccivil_03/_Ato2011-2014/2014/Decreto/D8262.htm) (in Portuguese).

21. Non-smokers' Health Act 1985. Ottawa (ON): Government of Canada; last amended 17 October 2018 (https://laws-lois.justice.gc.ca/eng/acts/N-23.6/page-1.html).

22. Cabin operations safety best practices guide, 6th edition. Montreal (QC): International Air Transport Association (IATA); 2016.

23. Use of e-cigarettes in public places and workplaces: advice to inform evidence-based policy making. London: Public Health England Publications; 2016 (https://assets.publishing.service.gov.uk/government/uploads/system/uploads/attachment_data/file/768952/PHE-advice-on-use-of-e-cigarettes-in-public-places-and-workplaces.PDF).

24. Industry guide on mandatory reporting under the Canada Consumer Product Safety Act – Section 14 "Duties in the event of an incident". Ottawa (ON): Government of Canada; 2018 (https://www.canada.ca/en/health-canada/services/consumer-product-safety/legislation-guidelines/acts-regulations/canada-consumer-product-safety-act/industry/guide-mandatory-reporting-section-14.html).

25. Guidance on preparing and maintaining documents under the Canada Consumer Product Safety Act (CCPSA) – Section 13. Ottawa (ON): Government of Canada; 2011 (amended 2012) (https://www.canada.ca/en/health-canada/services/consumer-product-safety/legislation-guidelines/guidelines-policies/guidance-preparing-maintaining-documents-under-canada-consumer-product-safety-act-section-13.html).

26. What is CISS? In: Korea Consumer Agency: Consumer Injury Surveillance System [website]. Chungbuk: Korea Consumer Agency; 2020 (https://www.ciss.go.kr/english/contents.do?key=595).

Annex 2   Additional regulatory requirements by the provincial jurisdictions of Canada

This annex shows the specific regulation of electronic nicotine and non-nicotine delivery systems (EN&NNDS), by Canadian province.

Province	Applicable law	Main provisions
British Columbia	Tobacco and Vapour Products Control Act. Date in effect: 1 September 2016 (1)	Under this law, the following are banned: • EN&NNDS use in all enclosed public spaces, including all public and private school grounds, workplaces and health-care facilities other than in designated areas; • sale and supply to minors (under 19); • sales wherever tobacco sales are banned; • any promotion in stores except at point of sale (POS) showing availability and price, including duty-free shops; - all POS displays except at POS where minors are not permitted to enter; and • vending machines in adult-only venues, including duty-free shops.
Manitoba	The Non-Smokers Health Protection and Vapour Products Act. Royal assent received: 5 November 2015 (2)	Under this law, the following are banned: • EN&NNDS use in enclosed public places and other places where smoking is presently prohibited, including workplaces, and work vehicles with more than one occupant (the following places are exempted from the ban to use EN&NNDS: where EN&NNDS are predominately sold and in designated smoking/vaping rooms in hotels and group living facilities); • EN&NNDS advertising and promotion as applicable to tobacco products; and • sale and supply to minors (under 18).
New Brunswick	Smoke-free Places Act, 2011. Tobacco and Electronic Cigarette Sales Act. As amended, date in effect: 10 November 2018 (3)	Provisions of the Act include the following bans: • EN&NNDS use where smoking is prohibited, including enclosed public spaces, workplaces, restaurants and bars, and vehicles when an individual under the age of 16 is present (there are exemptions for some hotel rooms and private residences); • sale and supply to minors (under 19); • sales of EN&NNDS wherever tobacco sales are banned; and • indoor and outdoor advertising and promotional materials, even within vape shops.
Newfoundland and Labrador	Smoke-Free Environment Act (as amended) (4), Tobacco and Vapour Products Control Act (as amended) (5). Date in effect: 17 October 2018	The following are banned: • sale to minors; • sale wherever tobacco sales are banned; • POS promotion, products and promotional materials (cannot be visible inside or outside the shop); - restrictions on signage inside shops; and • vape shops are allowed to operate if the only business conducted is the sale of vaping products. Treatment: treats EN&NNDS the same as combustible tobacco cigarettes, prohibiting the use of EN&NNDS in any place to which the public customarily has access, including workplaces, private clubs, licensed restaurants, bus shelters, and health-care and educational facilities. Young people: prohibits the use of EN&NNDS in motor vehicles when occupied by a person under the age of 16 and prohibits the sale of EN&NNDS and other vaping products to persons under the age of 19. Communications and advertising: there currently are no restrictions on EN&NNDS advertising. Vape shops can provide consumers with testimonials and health information regarding vaping.

Annex 2 Additional regulatory requirements by the provincial jurisdictions of Canada

Continued

Province	Applicable law	Main provisions
Nova Scotia	Smoke-free Places Act (amended) (6) and Tobacco Access Act (amended) (7). Date in effect: 31 May 2015	The following are banned: EN&NNDS use where smoking is prohibited, including enclosed public spaces, workplaces, education and • recreation facilities, restaurants and bars, and vehicles when an individual under the age of 19 is present; - sale and supply to minors (under 19 years); • possession by minors; • sale in pharmacies; • requirement to display age-restriction signage; and • POS promotion, except inside vape shops.
Ontario	Electronic Cigarettes Act, 2015, replaced by Smoke-Free Ontario Act, 2017 (last amended 2019) (8)	The following are banned: • EN&NNDS use where smoking is prohibited, including enclosed public place and workplaces, group • living facilities, public vehicles or vehicles used in the course of employment while carrying two or more employees (there are exceptions for guest rooms in hotels and designated areas in residential facilities); - sale and supply to minors (under 19); • supply or offer to supply health-care facilities and pharmacies; • sale wherever tobacco sales are banned (regulatory authority to make some exemptions); • promotion outdoors, at POS (only signage showing price and availability allowed), and premises to which children are permitted access; and • display visible to children in any place or premises in which tobacco or tobacco-related products are sold.
Quebec	Tobacco Control Act. Date in effect: 26 November 2015 (9)	Under the law, the following are banned: • EN&NNDS use where smoking is prohibited, including enclosed public spaces, workplaces, educational and recreational facilities, restaurants and bars, and vehicles when an individual under the age of 16 is present; - sale and supply to minors; • sale wherever tobacco sales are banned; and • outdoor signage restricted, shops are only allowed to show availability and price. All prohibitions that apply to tobacco promotion also apply to EN&NNDS. The ban on flavours in cigarettes does not apply to e-liquids.
Prince Edward Island	Tobacco and Electronic Smoking Device Sales and Access Act (10) and regulations (11); in effect 1 October 2015	The following are banned: • EN&NNDS use where smoking is prohibited, including enclosed public spaces, workplaces, education and recreation facilities, restaurants and bars, and vehicles when an individual under the age of 19 is present; • sale and supply to minors and purchase by minors under the age of 19; • sale where tobacco sale is banned; • POS promotion and promotion visible from outside retail premises; • vape shops can only display e-cigarettes if individuals under the age of 19 are not permitted; and • any advertising that is misleading regarding the characteristics, health effects and health hazards of these devices.

Reference[6]

1. Tobacco and Vapour Products Control Act, RSBC 1996, c 451. Victoria (BC): Legislative Assembly of British Columbia; 1996 (amended 2018) (http://canlii.ca/t/53gn3).
2. The Non-Smokers Health Protection and Vapour Products Act. Winnipeg (MB): Legislative Assembly of Manitoba; 1990 (amended 2019) (http://canlii.ca/t/53ncl).
3. Smoke-free Places Act, RSNB 2011, c 222. Fredericton (NB): Legislative Assembly of New Brunswick; 2011 (amended 2019) (http://canlii.ca/t/53l69).
4. Smoke-free Environment Act, 2005, SNL 2005, c S-16.2 (as amended). St John's (NL): Newfoundland and Labrador House of Assembly; 2018 (http://canlii.ca/t/53gdz).
5. Tobacco and Vapour Products Control Act, SNL 1993, c T-4.1. St John's (NL): Newfoundland and Labrador House of Assembly; 1993 (amended 2018) (http://canlii.ca/t/53gf9).
6. Smoke-free Places Act (amended). Chapter 54 of the Acts of 2007. Halifax (NS): Nova Scotia Legislature; 2007 (https://nslegislature.ca/legc/bills/60th_2nd/3rd_read/b006.htm).
7. Tobacco Access Act, SNS 1993, c 14. Chapter 14 of the Acts of 1993. Halifax (NS): Nova Scotia Legislature; 1993 (https://www.canlii.org/en/ns/laws/stat/sns-1993-c-14/latest/sns-1993-c-14.html).
8. Smoke-Free Ontario Act, 2017, SO 2017, c 26, Sch 3. Toronto (ON): Legislative Assembly of Ontario: 2017 (http://canlii.ca/t/53m3f).
9. Tobacco Control Act, 2015, c. 28, s. 1. Quebec City (QB): National Assembly of Quebec; 2015 (http://legisquebec.gouv.qc.ca/en/ShowDoc/cs/L-6.2).
10. Tobacco and Electronic Smoking Device Sales, RSPEI 1988, c T-3.1. Charlottetown: Prince Edward Island Government; amended 2020 (http://canlii.ca/t/548v9).
11. Tobacco and Electronic Smoking Device Sales and Access Regulations, PEI Reg EC414/05. Charlottetown: Prince Edward Island Government; 2020 (http://canlii.ca/t/548vb).12.

[6] All weblinks accessed 18 March 2020.